LEARN TO ADD AND SUBTRACT WITH A JAPANESE ABACUS

Author: Paul Green

CONTENTS

BASIC TERMINOLOGY

Note: It is not necessary to memorise all of this, it is just for reference .

- Addition: Finding the sum or total, for example 8+7=15 is an addition
- Decimal: Figures placed to the right of a decimal point. Example the 8 of 0.8
- Decimal fraction: A fraction that is written as a decimal. Example 0.46
- Decimal point: The point between a whole number and a decimal fraction. Example 9.86 (the point between the 9 and the 8)
- Decimal: Any number used in the decimal system, specifically numbers that have one or more digits to the right of the decimal point. Example 43.7 or 0.68
- Decimal system: A number system based on units of 10
- Digit: A symbol used to show a number. Example 6
- One-hundredth: If you divide the number 100 into 100 parts, one of those parts is a hundredth. Example 1/100 or 0.01
- One-tenth: If you divide the number 10 into 10 parts, one of those parts is a tenth. Example 1/10 or 0.1
- One-thousandth: If you divide the number 1000 into 1000 parts, one of those parts is a thousandth. Example 1/1000 or 0.001
- Subtraction: Taking one number away from another, for example 8-3=5 is a subtraction
- Whole number: A counting number from 0 to infinity. Examples 0, 1, 2, 3, 4, 5 etc. Note: 20 consists of two whole numbers

INTRODUCTION

The Japanese abacus is also called the Soroban.

This is abacus written in Japanese そろばん

The Japanese abacus is mostly used for adding and subtracting numbers.

The Japanese abacus has a wooden frame and five beads per column, one bead above the beam (worth 5) and four beads below the beam (worth 1 each).

The beads that are pushed against the frame are called 'Unregistered' beads and the beads that are pushed against the beam are called 'Registered' beads.

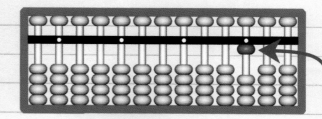

One bead registered
(One lower bead is pushed against the beam. All other beads are unregistered.)

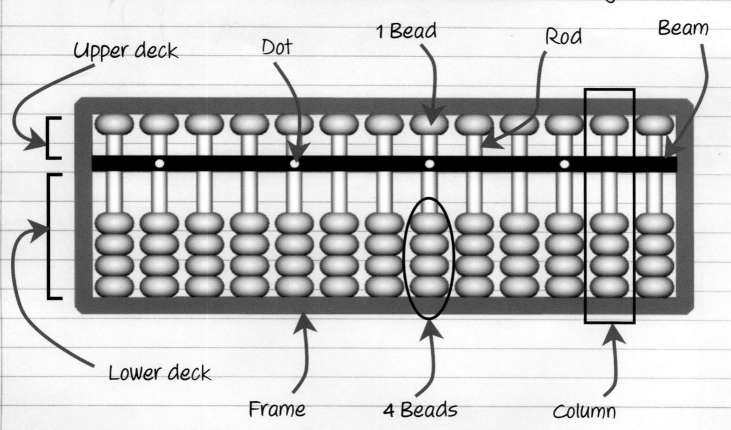

Upper deck   Dot   1 Bead   Rod   Beam

Lower deck

Frame   4 Beads   Column

The picture below shows the values of each column on the abacus.

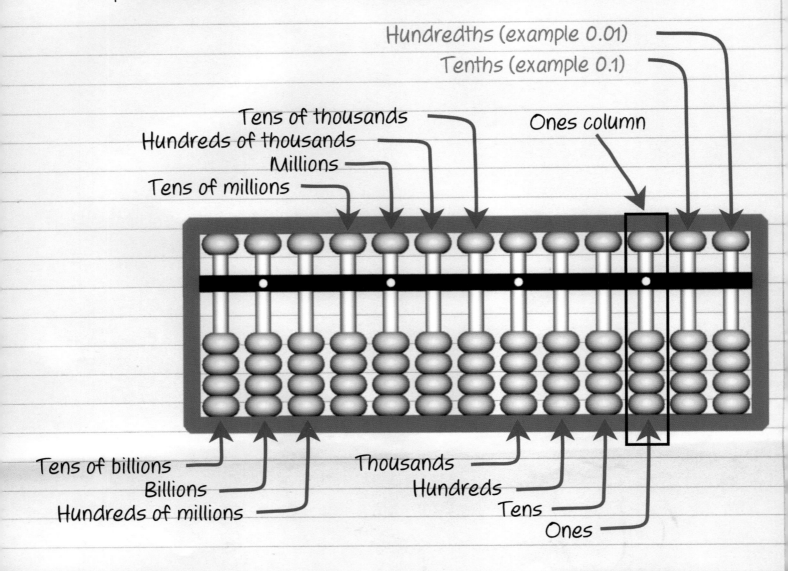

Hundredths (example 0.01)
Tenths (example 0.1)
Tens of thousands
Hundreds of thousands
Millions
Tens of millions
Ones column

Tens of billions
Billions
Hundreds of millions
Thousands
Hundreds
Tens
Ones

Note: The first two columns on the right side are for decimals ONLY, whole numbers start on the third column from the right side of the abacus.

With the Japanese abacus it is possible to perform decimal (base-10) calculations. The decimal system is based on positioning digits in their places. The value of each bead increases ten times with each shift to the left.

This digit is 10 times larger than this digit

1111

MOVING THE BEADS

What fingers do you use to move the beads?

We will use the thumb and index finger. Some people like to use the thumb, index and middle finger but we will keep it simple for this book.

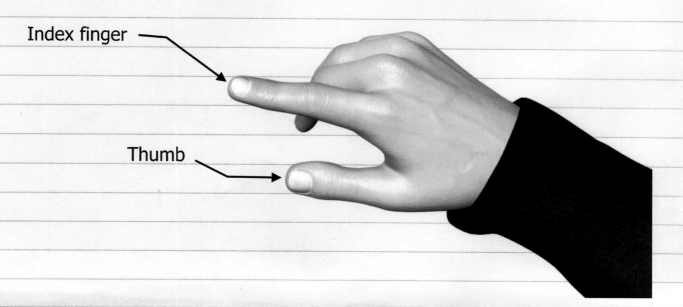

Index finger

Thumb

Thumb

Used to register the LOWER beads (to push them towards the beam).

Index finger

Used to register the UPPER beads (to push them towards the beam).
Used to unregister (away from the beam) ALL beads.

Bead order

When registering and unregistering a number, always start with the highest value column first, then work towards the lowest value column. For example, when registering the number 23 start by registering the 2 (20) and then the 3 (3).

On the next pages there are pictures which show the finger movements for registering and unregistering.

Using the thumb

Register 1 lower bead

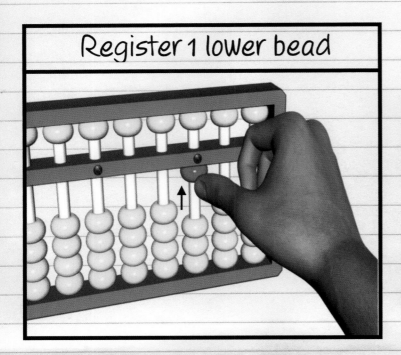

Register 2 lower beads

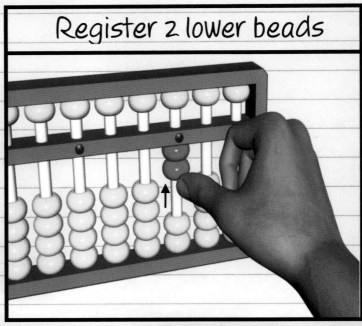

Register 3 lower beads

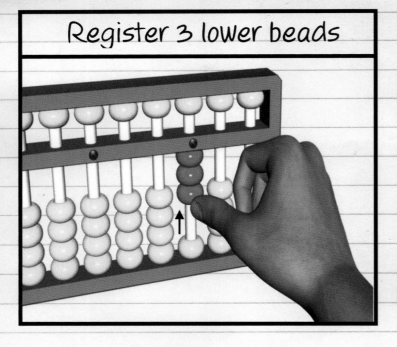

Register 4 lower beads

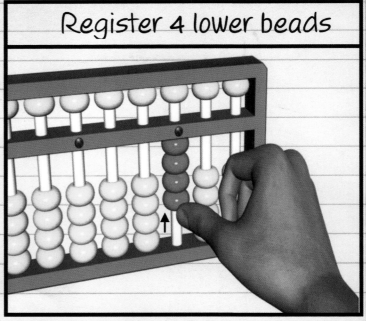

Unregister 1 lower bead

Unregister 2 lower beads

Unregister 3 lower beads

Unregister 4 lower beads

Unregister 1 upper bead

Register 1 upper bead

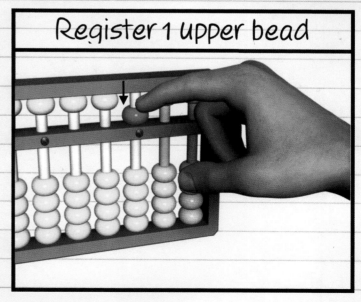

REGISTERING NUMBERS ON THE ABACUS

Registering is when you move a bead so that it is to be included in your calculation. When you move a bead so that it touches the beam, then this bead is said to be registered. When you move a bead away from the beam this is said to be unregistered, meaning that it will no longer be counted in your calculation. Throughout the book the computer terms 'Register' and 'Unregister' will be used to refer to beads that are moved towards the beam (register a bead) or away from the beam (unregister a bead).

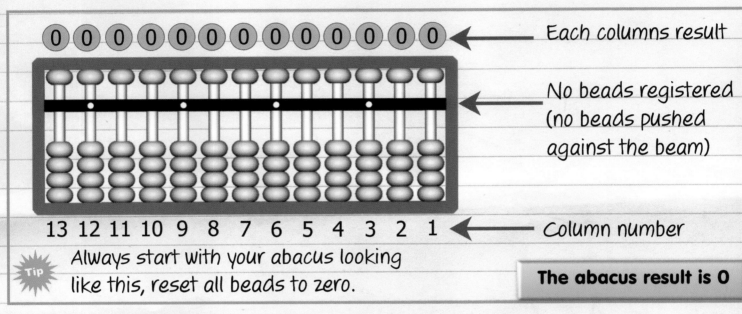

Each columns result

No beads registered (no beads pushed against the beam)

13 12 11 10 9 8 7 6 5 4 3 2 1 ← Column number

Tip Always start with your abacus looking like this, reset all beads to zero.

The abacus result is 0

Register 1 on the abacus

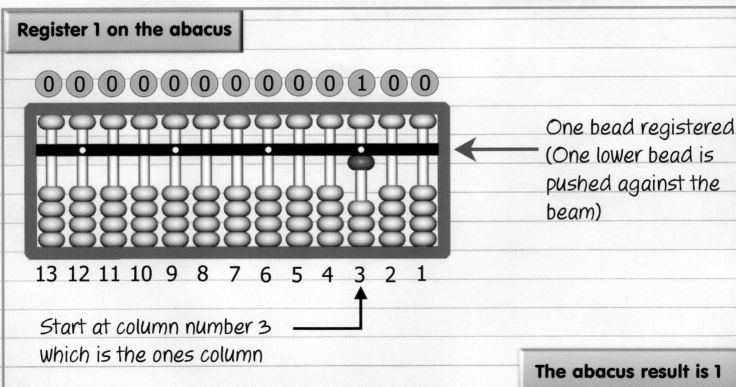

One bead registered (One lower bead is pushed against the beam)

13 12 11 10 9 8 7 6 5 4 3 2 1

Start at column number 3 which is the ones column

The abacus result is 1

Register 5 on the abacus

Remember the upper bead is worth 5.

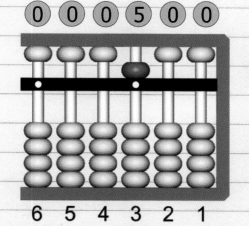

One bead registered
(One upper bead is pushed
against the beam)

6 5 4 3 2 1

The abacus result is 5

Register 6 on the abacus

The upper bead is worth 5, the lower bead is worth 1.

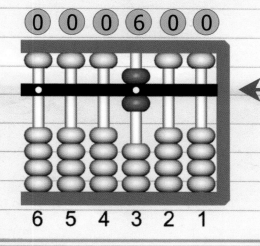

Two beads registered
(One upper bead and one
lower bead are pushed
against the beam)

6 5 4 3 2 1

The abacus result is 6

Register 8 on the abacus

The upper bead is worth 5 plus 3 lower beads
worth 1 each. This gives a total of 8.

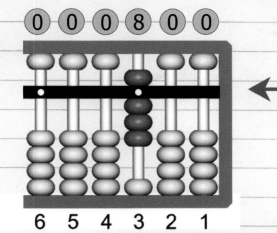

Four beads registered
(One upper bead and
three lower beads are
pushed against the beam)

6 5 4 3 2 1

The abacus result is 8

HOW TO REGISTER MULTI-DIGIT NUMBERS

A multi-digit number is any number that has more than one digit.
Let's start with a 5 digit number.

23456

First register the '2' digit (leftmost) in the 7th column.
Why the 7th column? Because the number has 5 digits and we are not using the first 2 columns.

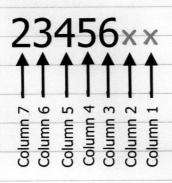

23456 × ×

Column 7
Column 6
Column 5
Column 4
Column 3
Column 2
Column 1

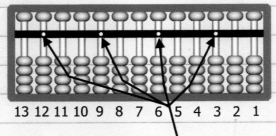

13 12 11 10 9 8 7 6 5 4 3 2 1

Tip To find the column number quickly, look at the dots. The dots are every 3rd column, 3, 6, 9, and 12.

Register 23456

Column 6 dot

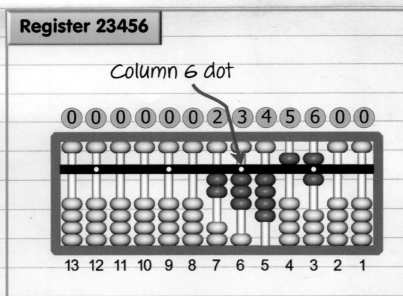

0 0 0 0 0 0 2 3 4 5 6 0 0

13 12 11 10 9 8 7 6 5 4 3 2 1

- We will put 23456 on the abacus
- 23456 has 5 digits, so use 5 columns (start on column 7)

- Column 7, register 2 lower beads
- Column 6, register 3 lower beads
- Column 5, register 4 lower beads
- Column 4, register 1 upper bead
- Column 3, register 1 upper bead and 1 lower bead
(total on this column is 5+1=6)

The abacus result is 23456

Things to remember before we move on:
- Don't use columns 1 and 2 (keep those for decimal numbers)
- The total digits of the whole number plus 2 = the column where we start to register our number
- The dots help us find the column number

Register 15

Start with the leftmost digit first (in this example the 1 of the 15) then move to the right to register the next digit.

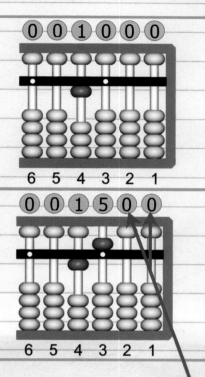

① ① ① ① ① ①
6 5 4 3 2 1

First register the '1' digit (leftmost) in the 'tens column', column 4.

15

① ① ① ⑤ ① ①
6 5 4 3 2 1

Second register the '5' digit

15

The two zeros after the number 15 are for decimal numbers, tenths (example the digit 3 in 0.3) for column 2 and hundredths in column 1 (example the digit 6 in 0.46).

The abacus result is 15

Register 316

Column 6 dot

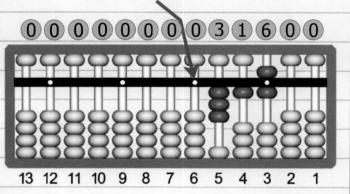

① ① ① ① ① ① ① ① ③ ① ⑥ ① ①
13 12 11 10 9 8 7 6 5 4 3 2 1

- We will put 316 on the abacus
- 316 has 3 digits, so use 3 columns (start on column 5)
- Column 5, register 3 lower beads
- Column 4, register 1 lower bead
- Column 3, register 1 upper bead and 1 lower bead

The abacus result is 316

Register 556677

Column 9 dot

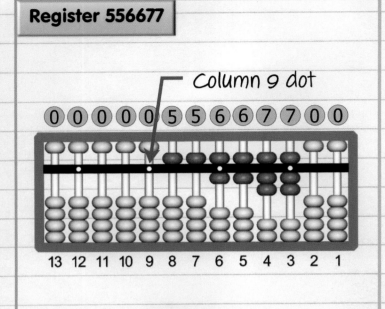

0 0 0 0 0 5 5 6 6 7 7 0 0

13 12 11 10 9 8 7 6 5 4 3 2 1

- We will put 556677 on the abacus
- 556677 has 6 digits, so use 6 columns (start on column 8)
- Column 8, register 1 upper bead
- Column 7, register 1 upper bead
- Column 6, register 1 upper bead and 1 lower bead
- Column 5, register 1 upper bead and 1 lower bead
- Column 4, register 1 upper bead and 2 lower beads
- Column 3, register 1 upper bead and 2 lower beads

The abacus result is 556677

Register 920030598

Column 12 dot

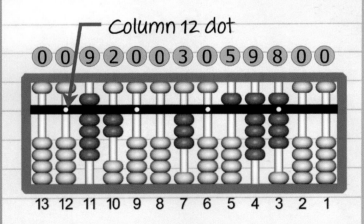

0 0 9 2 0 0 3 0 5 9 8 0 0

13 12 11 10 9 8 7 6 5 4 3 2 1

Use dot 12 to find column 11

- We will put 920030598 on the abacus
- 920030598 has 9 digits, so use 9 columns (start on column 11)
- Column 11, register 1 upper bead and 4 lower beads
- Column 10, register 2 lower beads
- Column 9, do nothing
- Column 8, do nothing
- Column 7, register 3 lower beads
- Column 6, do nothing
- Column 5, register 1 upper bead
- Column 4, register 1 upper bead and 4 lower beads
- Column 3, register 1 upper bead and 3 lower beads

The abacus result is 920030598

ADDITION

Addition is adding numbers to get the sum of those numbers

Addition - things to remember:

Register your numbers from left to right , for example for number 31 register the 3 first, and 1 last.

Each digit must be registered in the correct column , for example with 31 the 3 for column 4 (tens column) and the 1 for column 3 (ones column).

Example: 31 + 13

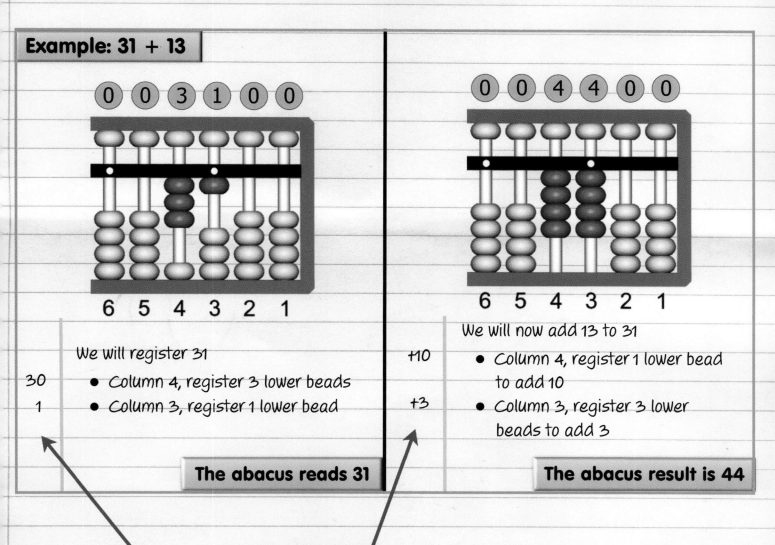

We will register 31

30   • Column 4, register 3 lower beads

1   • Column 3, register 1 lower bead

The abacus reads 31

We will now add 13 to 31

+10   • Column 4, register 1 lower bead to add 10

+3   • Column 3, register 3 lower beads to add 3

The abacus result is 44

These columns are useful to see the amount that you are adding.

For example:

30 means that you have just registered 30

+3 means that you have just added 3

Example: 27 + 62

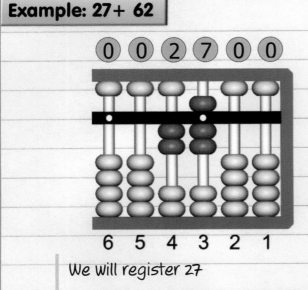

We will register 27

20   • Column 4, register 2 lower beads
      (this is the 2 of the 27)

7   • Column 3, register 1 upper bead
      and 2 lower beads
      (this is the 7 of the 27)

The abacus reads 27

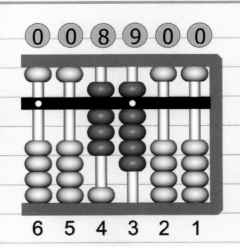

We will now add 62 to 27

+60   • Column 4, register 1 upper bead
      and 1 lower bead to add 60
      (Total 50+10=60) remember,
      column 4 beads are in the tens column

+2   • Column 3, register 2 lower beads
      to add 2

The abacus result is 89

Example: 527 + 462

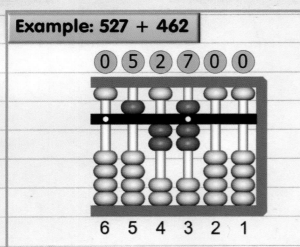

We will register 527

500   • Column 5, register 1 upper bead
      (this is the 5 of the 527)

20   • Column 4, register 2 lower beads
      (this is the 2 of the 527)

7   • Column 3, register 1 upper bead
      and 2 lower beads (5+2=7)
      (this is the 7 of the 527)

The abacus reads 527

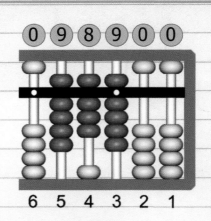

We will now add 462 to 527

+400 • Column 5, register 4 lower beads to add
      400

+60 • Column 4, register 1 upper and 1
      lower bead to add 60
      (Total 50+10=60)

+2 • Column 3, register 2 lower beads to add 2

The abacus result is 989

Example: 3213 + 1220

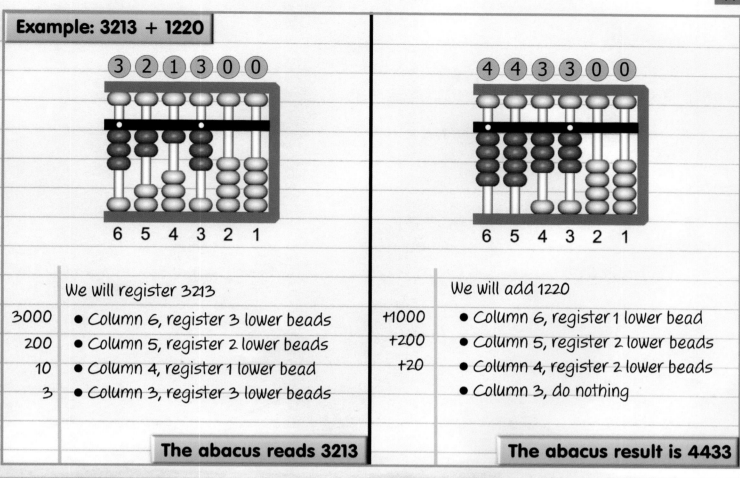

We will register 3213

3000	● Column 6, register 3 lower beads
200	● Column 5, register 2 lower beads
10	● Column 4, register 1 lower bead
3	● Column 3, register 3 lower beads

The abacus reads 3213

We will add 1220

+1000	● Column 6, register 1 lower bead
+200	● Column 5, register 2 lower beads
+20	● Column 4, register 2 lower beads
	● Column 3, do nothing

The abacus result is 4433

REGISTER AND UNREGISTER IN THE SAME COLUMN

Sometimes we need to unregister and register in the same column.
For example if we need to add 3 beads to an already registered 4 lower beads to make 7, we need to register 1 upper and unregister 2 lower beads (+5-2=3).

Example: 74 + 13

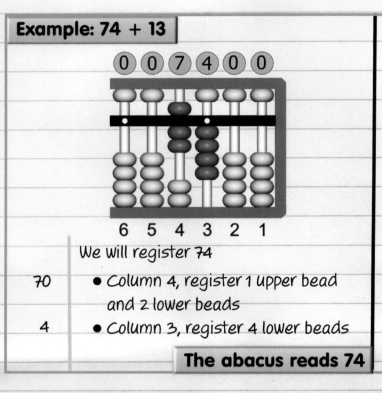

We will register 74

| 70 | • Column 4, register 1 upper bead and 2 lower beads |
| 4 | • Column 3, register 4 lower beads |

The abacus reads 74

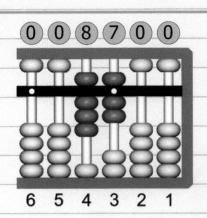

We will now add 13 to 74

| +10 | • Column 4, register 1 lower bead |
| +3 | • Column 3, register 1 upper bead (+5) and unregister 2 lower beads (-2) |

The abacus result is 87

Example: 84 + 12

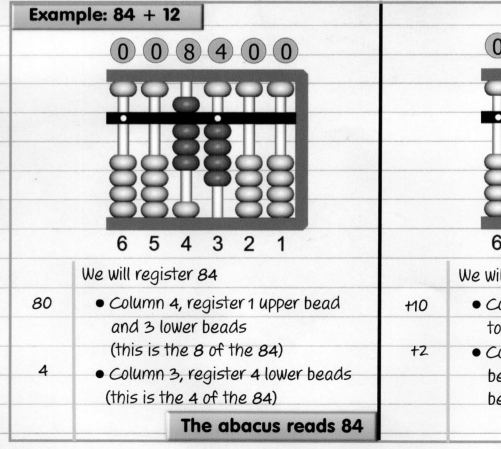

We will register 84

| 80 | • Column 4, register 1 upper bead and 3 lower beads (this is the 8 of the 84) |
| 4 | • Column 3, register 4 lower beads (this is the 4 of the 84) |

The abacus reads 84

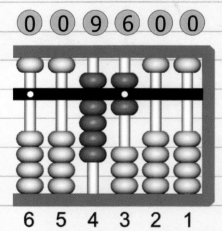

We will now add 12 to 84

| +10 | • Column 4, register 1 lower bead to add 10 |
| +2 | • Column 3, register 1 upper bead (+5) and unregister 3 lower beads (-3) |

The abacus result is 96

NOT ENOUGH BEADS

When you **don't have enough beads,** move to the next LEFT column to help.

For example, when you try to add 4 to the already registered number 8, you don't have enough beads in the column to do it.

You can only register a maximum of 9 in each column (4 lower beads and 1 upper bead, 4+5=9).

When this happens, we need to use the **'Not enough beads list'**.

1=10-9
2=10-8
3=10-7
4=10-6
5=10-5
6=10-4
7=10-3
8=10-2
9=10-1

Let's say we need to add 3 to a column but we don't have enough beads.

From the list **3=10-7**

10 is the number to **register,** in the next LEFT column (1 lower bead).

7 is the number to **unregister** in our column.

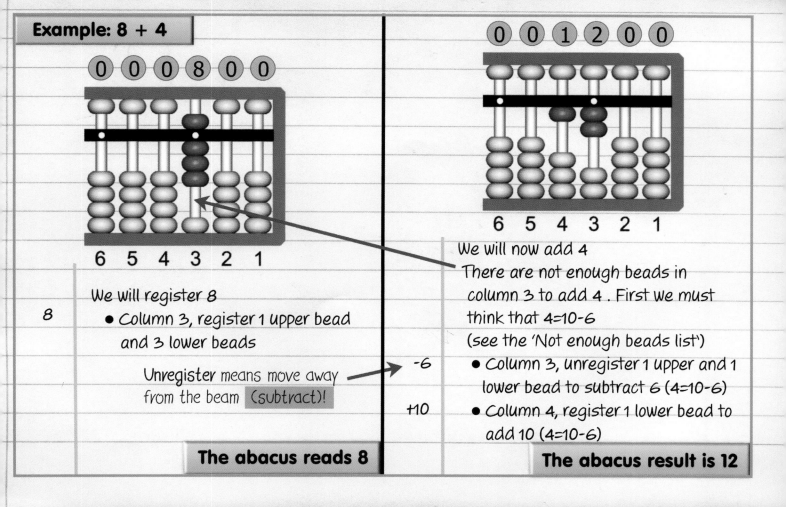

Example: 8 + 4

0 0 0 8 0 0

6 5 4 3 2 1

8

We will register 8
- Column 3, register 1 upper bead and 3 lower beads

Unregister means move away from the beam **(subtract)!**

The abacus reads 8

0 0 1 2 0 0

6 5 4 3 2 1

We will now add 4
There are not enough beads in column 3 to add 4. First we must think that 4=10-6
(see the 'Not enough beads list')

-6
- Column 3, unregister 1 upper and 1 lower bead to subtract 6 (4=10-6)

+10
- Column 4, register 1 lower bead to add 10 (4=10-6)

The abacus result is 12

Example: 9 + 5

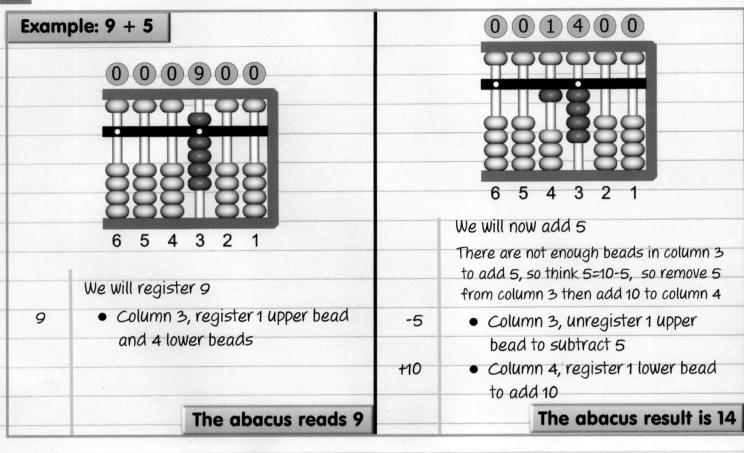

We will register 9

9
- Column 3, register 1 upper bead and 4 lower beads

The abacus reads 9

We will now add 5

There are not enough beads in column 3 to add 5, so think 5=10-5, so remove 5 from column 3 then add 10 to column 4

-5
- Column 3, unregister 1 upper bead to subtract 5

+10
- Column 4, register 1 lower bead to add 10

The abacus result is 14

Example: 39 + 53

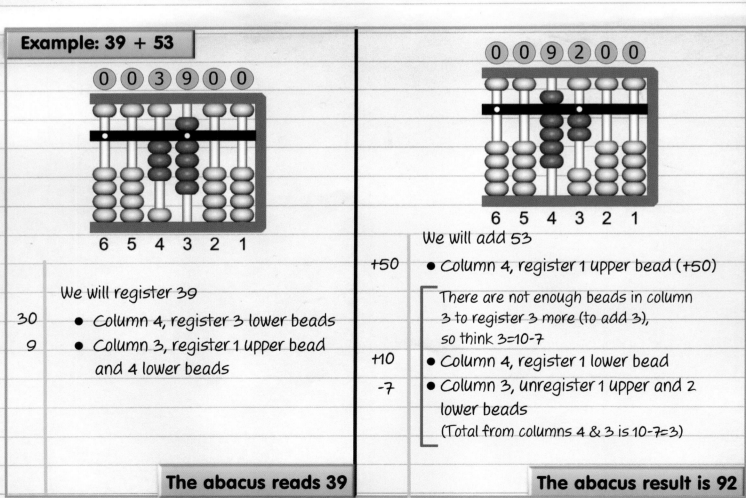

We will register 39

30
- Column 4, register 3 lower beads

9
- Column 3, register 1 upper bead and 4 lower beads

The abacus reads 39

We will add 53

+50
- Column 4, register 1 upper bead (+50)

There are not enough beads in column 3 to register 3 more (to add 3), so think 3=10-7

+10
- Column 4, register 1 lower bead

-7
- Column 3, unregister 1 upper and 2 lower beads

(Total from columns 4 & 3 is 10-7=3)

The abacus result is 92

SKIPPED COLUMNS

Sometimes we have to SKIP a column. See examples below.

When a column doesn't have enough beads left on it to make the addition, we move to the next LEFT column to help. Sometimes the next left column also doesn't have enough beads on it, so we SKIP this column and move again to the next left column until you reach a column that has enough beads to use.
See below how it works.

- SKIP a column when there are not enough beads to use in that column

- We will see this symbol ← when we need to skip a column (move on to the next left column)

- UNREGISTER all beads in any skipped columns

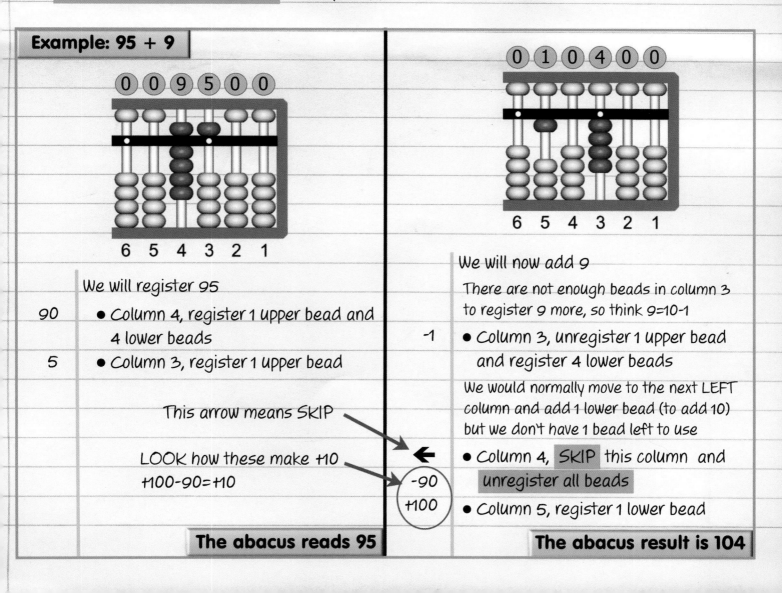

Example: 95 + 9

We will register 95
- 90 · Column 4, register 1 upper bead and 4 lower beads
- 5 · Column 3, register 1 upper bead

This arrow means SKIP

LOOK how these make +10
+100-90=+10

The abacus reads 95

We will now add 9
There are not enough beads in column 3 to register 9 more, so think 9=10-1
- -1 · Column 3, unregister 1 upper bead and register 4 lower beads

We would normally move to the next LEFT column and add 1 lower bead (to add 10) but we don't have 1 bead left to use
- · Column 4, SKIP this column and unregister all beads
- -90 / +100 · Column 5, register 1 lower bead

The abacus result is 104

Example: 995 + 9

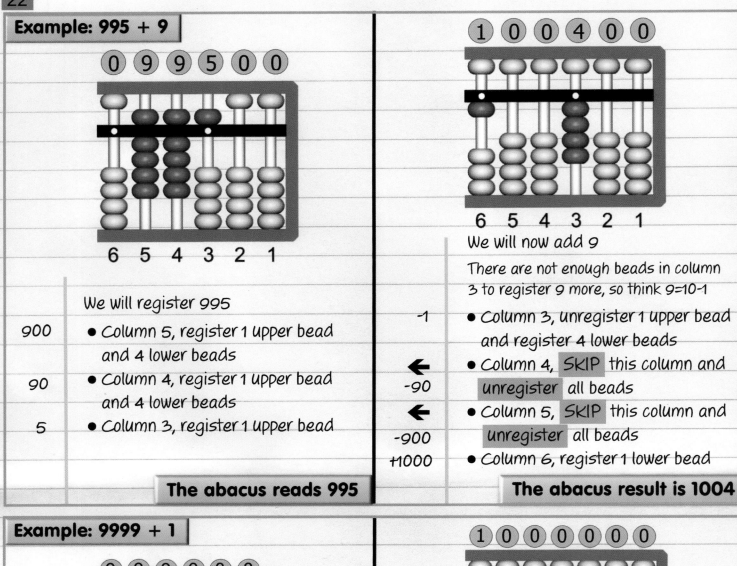

⓪ ⑨ ⑨ ⑤ ⓪ ⓪

6 5 4 3 2 1

We will register 995

900	• Column 5, register 1 upper bead and 4 lower beads
90	• Column 4, register 1 upper bead and 4 lower beads
5	• Column 3, register 1 upper bead

The abacus reads 995

① ⓪ ⓪ ④ ⓪ ⓪

6 5 4 3 2 1

We will now add 9

There are not enough beads in column 3 to register 9 more, so think 9=10-1

-1	• Column 3, unregister 1 upper bead and register 4 lower beads
← -90	• Column 4, SKIP this column and unregister all beads
← -900	• Column 5, SKIP this column and unregister all beads
+1000	• Column 6, register 1 lower bead

The abacus result is 1004

Example: 9999 + 1

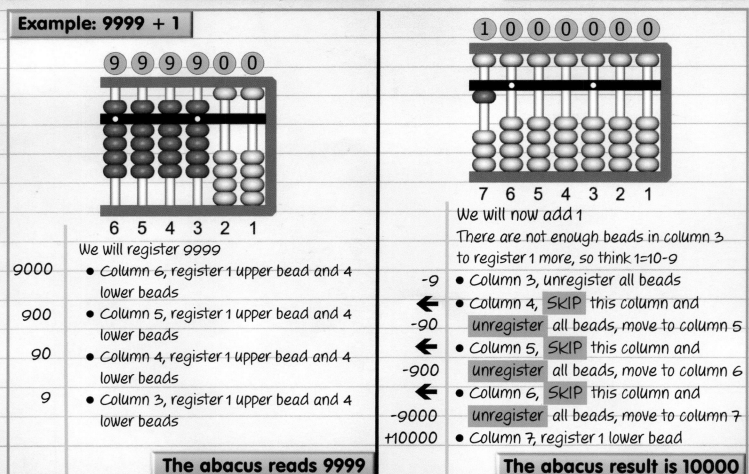

⑨ ⑨ ⑨ ⑨ ⓪ ⓪

6 5 4 3 2 1

We will register 9999

9000	• Column 6, register 1 upper bead and 4 lower beads
900	• Column 5, register 1 upper bead and 4 lower beads
90	• Column 4, register 1 upper bead and 4 lower beads
9	• Column 3, register 1 upper bead and 4 lower beads

The abacus reads 9999

① ⓪ ⓪ ⓪ ⓪ ⓪ ⓪

7 6 5 4 3 2 1

We will now add 1

There are not enough beads in column 3 to register 1 more, so think 1=10-9

-9	• Column 3, unregister all beads
← -90	• Column 4, SKIP this column and unregister all beads, move to column 5
← -900	• Column 5, SKIP this column and unregister all beads, move to column 6
← -9000	• Column 6, SKIP this column and unregister all beads, move to column 7
+10000	• Column 7, register 1 lower bead

The abacus result is 10000

When we want to add more than two numbers on the abacus, just find the sum of the first two , then add the next number to that sum .

Keep adding one number to the sum of the previous numbers until all the numbers have been added.

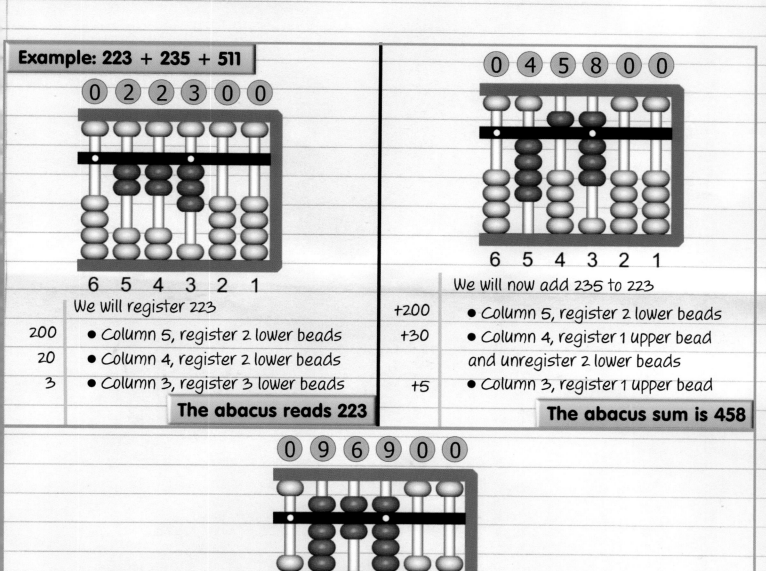

Example: 223 + 235 + 511

(0)(2)(2)(3)(0)(0)

6 5 4 3 2 1

We will register 223

200	• Column 5, register 2 lower beads
20	• Column 4, register 2 lower beads
3	• Column 3, register 3 lower beads

The abacus reads 223

(0)(4)(5)(8)(0)(0)

6 5 4 3 2 1

We will now add 235 to 223

+200	• Column 5, register 2 lower beads
+30	• Column 4, register 1 upper bead and unregister 2 lower beads
+5	• Column 3, register 1 upper bead

The abacus sum is 458

(0)(9)(6)(9)(0)(0)

6 5 4 3 2 1

We will now add 511 to the sum 458

+500	• Column 5, register 1 upper bead
+10	• Column 4, register 1 lower bead
+1	• Column 3, register 1 lower bead

The abacus result is 969

Example: 525631 + 253160 + 1210

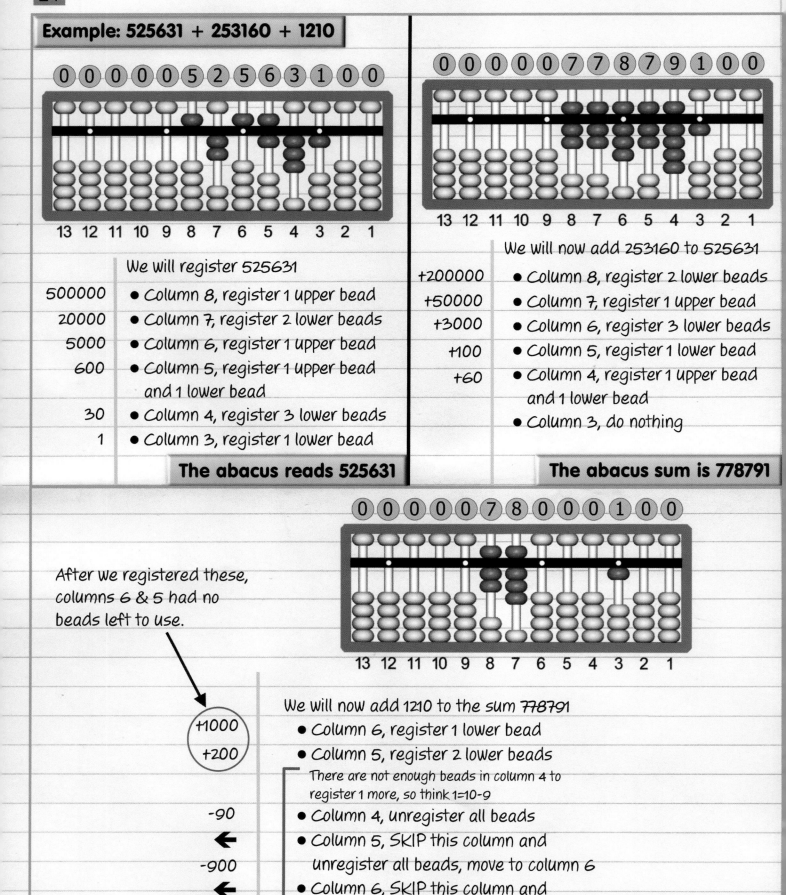

0 0 0 0 0 5 2 5 6 3 1 0 0

13 12 11 10 9 8 7 6 5 4 3 2 1

We will register 525631

500000	● Column 8, register 1 upper bead
20000	● Column 7, register 2 lower beads
5000	● Column 6, register 1 upper bead
600	● Column 5, register 1 upper bead and 1 lower bead
30	● Column 4, register 3 lower beads
1	● Column 3, register 1 lower bead

The abacus reads 525631

0 0 0 0 0 7 7 8 7 9 1 0 0

13 12 11 10 9 8 7 6 5 4 3 2 1

We will now add 253160 to 525631

+200000	● Column 8, register 2 lower beads
+50000	● Column 7, register 1 upper bead
+3000	● Column 6, register 3 lower beads
+100	● Column 5, register 1 lower bead
+60	● Column 4, register 1 upper bead and 1 lower bead
	● Column 3, do nothing

The abacus sum is 778791

0 0 0 0 0 7 8 0 0 0 1 0 0

13 12 11 10 9 8 7 6 5 4 3 2 1

After we registered these, columns 6 & 5 had no beads left to use.

We will now add 1210 to the sum 778791

+1000	● Column 6, register 1 lower bead
+200	● Column 5, register 2 lower beads
	There are not enough beads in column 4 to register 1 more, so think 1=10-9
-90	● Column 4, unregister all beads
←	● Column 5, SKIP this column and unregister all beads, move to column 6
-900	
←	● Column 6, SKIP this column and unregister all beads, move to column 7
-9000	
+10000	● Column 7, register 1 lower bead
	● Column 3, do nothing

The abacus result is 780001

ADDITION SUMMARY

- Register your numbers from left to right and in the correct column .

- Sometimes we need to unregister and register in the same column.

- When you don't have enough beads, move to the next LEFT column to help.

- If the next left column also doesn't have enough beads on it, unregister all beads in that column and SKIP to the next left column (you may need to unregister beads and skip more columns until you reach a column that has enough beads to use).

- When we want to add more than two numbers on the abacus, just find the sum of the first two , then add the next number to that sum .

SUBTRACTION

Subtraction is taking one number away from another to find the difference.

Subtraction - things to remember:

- Register your numbers from left to right, just the same as we did with addition, for example:

  for number 641 register the 6 first, 4 second and 1 last.

- Each digit must be registered in the correct column, for example with 641 the 6 is for column 5 (hundredths column), the 4 for column 4 (tens column) and the 1 for column 3 (ones column), just like we did with addition.

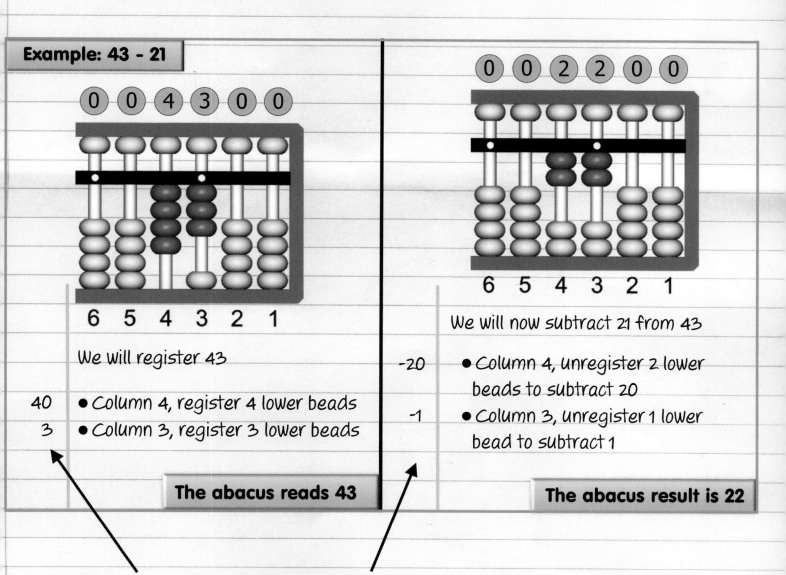

Example: 43 - 21

We will register 43

40 • Column 4, register 4 lower beads

3 • Column 3, register 3 lower beads

The abacus reads 43

We will now subtract 21 from 43

-20 • Column 4, unregister 2 lower beads to subtract 20

-1 • Column 3, unregister 1 lower bead to subtract 1

The abacus result is 22

These columns are useful to see the amount that you are subtracting.
For example:

40 means that you have just registered 40

-20 means that you have just subtracted 20

Example: 686 - 661

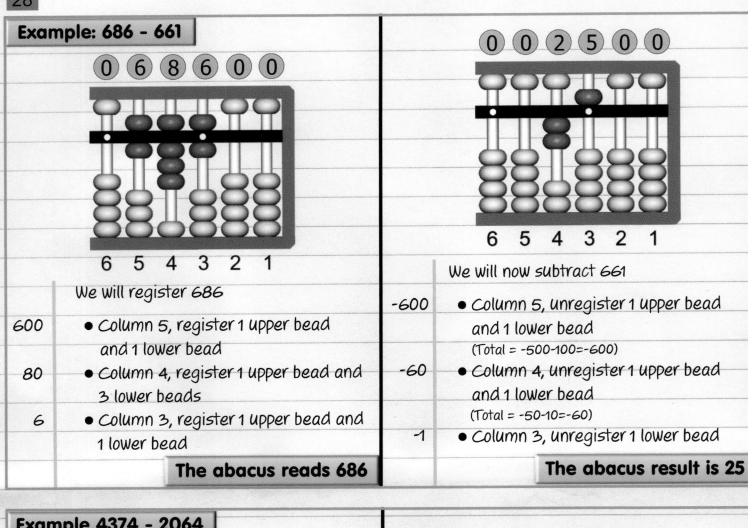

We will register 686

600	• Column 5, register 1 upper bead and 1 lower bead
80	• Column 4, register 1 upper bead and 3 lower beads
6	• Column 3, register 1 upper bead and 1 lower bead

The abacus reads 686

We will now subtract 661

-600	• Column 5, unregister 1 upper bead and 1 lower bead (Total = -500-100=-600)
-60	• Column 4, unregister 1 upper bead and 1 lower bead (Total = -50-10=-60)
-1	• Column 3, unregister 1 lower bead

The abacus result is 25

Example 4374 - 2064

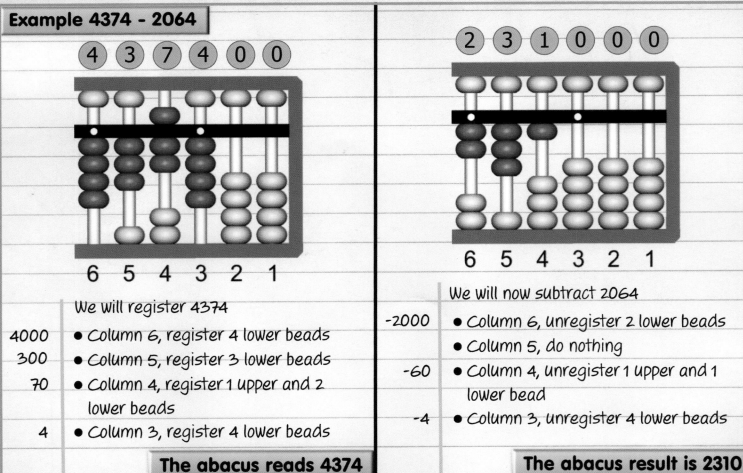

We will register 4374

4000	• Column 6, register 4 lower beads
300	• Column 5, register 3 lower beads
70	• Column 4, register 1 upper and 2 lower beads
4	• Column 3, register 4 lower beads

The abacus reads 4374

We will now subtract 2064

-2000	• Column 6, unregister 2 lower beads
	• Column 5, do nothing
-60	• Column 4, unregister 1 upper and 1 lower bead
-4	• Column 3, unregister 4 lower beads

The abacus result is 2310

SUBTRACTING WITH DIFFERENT AMOUNTS OF DIGITS

For example, when subtracting 234 - 21 We see that 234 has 3 digits and 21 only has 2.

Register the number that has the largest amount of digits, in this case it is 234.

Next, subtract the number with the smallest amount of digits, in this example the 21, from the largest digit number.

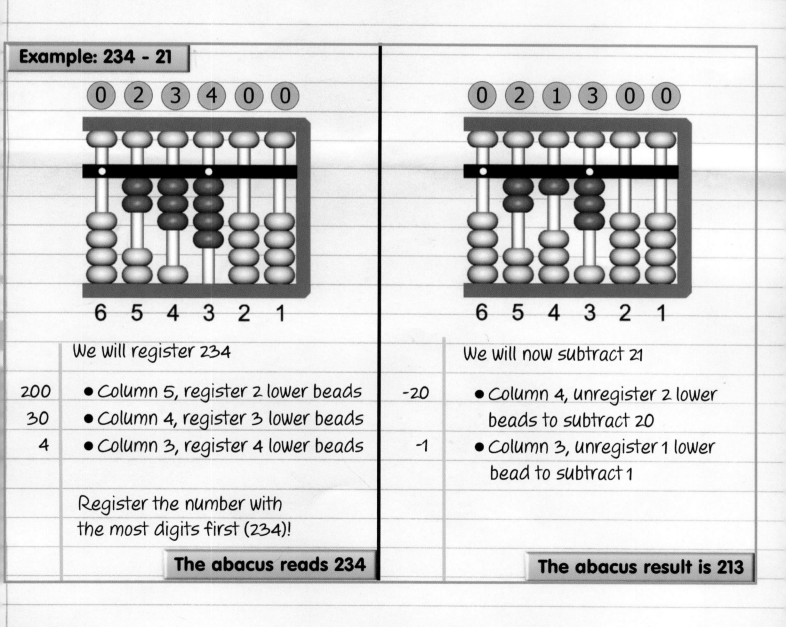

Example: 234 - 21

We will register 234

200	● Column 5, register 2 lower beads
30	● Column 4, register 3 lower beads
4	● Column 3, register 4 lower beads

Register the number with the most digits first (234)!

The abacus reads 234

We will now subtract 21

-20	● Column 4, unregister 2 lower beads to subtract 20
-1	● Column 3, unregister 1 lower bead to subtract 1

The abacus result is 213

Example: 6533 - 323

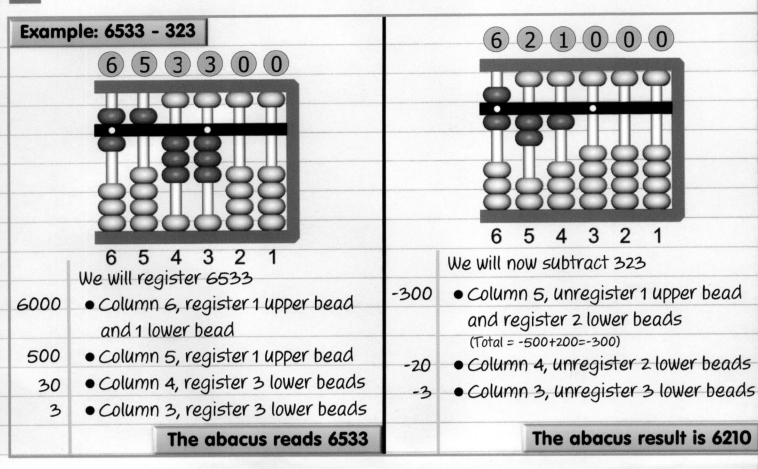

6 5 3 3 0 0

6 5 4 3 2 1

We will register 6533

6000	● Column 6, register 1 upper bead and 1 lower bead
500	● Column 5, register 1 upper bead
30	● Column 4, register 3 lower beads
3	● Column 3, register 3 lower beads

The abacus reads 6533

6 2 1 0 0 0

6 5 4 3 2 1

We will now subtract 323

-300	● Column 5, unregister 1 upper bead and register 2 lower beads
	(Total = -500+200=-300)
-20	● Column 4, unregister 2 lower beads
-3	● Column 3, unregister 3 lower beads

The abacus result is 6210

Example: 8543 - 432

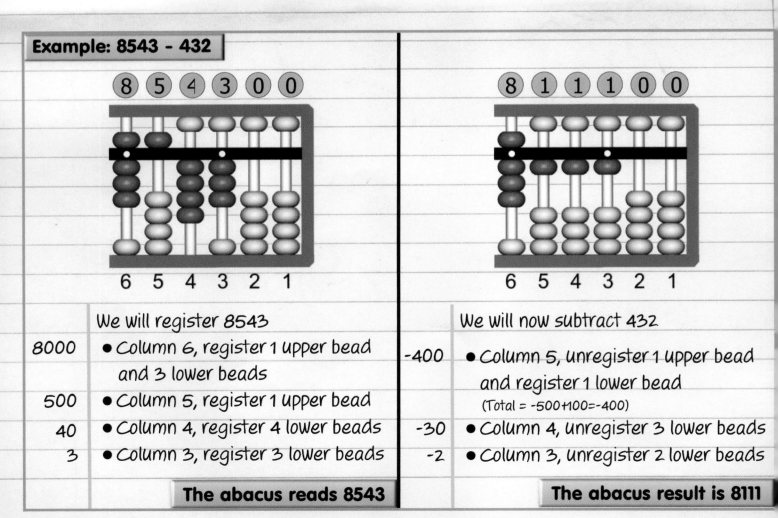

8 5 4 3 0 0

6 5 4 3 2 1

We will register 8543

8000	● Column 6, register 1 upper bead and 3 lower beads
500	● Column 5, register 1 upper bead
40	● Column 4, register 4 lower beads
3	● Column 3, register 3 lower beads

The abacus reads 8543

8 1 1 1 0 0

6 5 4 3 2 1

We will now subtract 432

-400	● Column 5, unregister 1 upper bead and register 1 lower bead
	(Total = -500+100=-400)
-30	● Column 4, unregister 3 lower beads
-2	● Column 3, unregister 2 lower beads

The abacus result is 8111

NOT ENOUGH BEADS

When you don't have enough beads, move to the next LEFT column to help.

For example, when you try to subtract 8 from an already registered number 12, you don't have enough beads in the column where the 2 of the 12 is, to do it. You can only unregister a maximum of 9 in each column (4 lower beads and 1 upper bead, -4-5=-9).

When this happens, we need to use the 'Not enough beads list for subtraction'.

-1=-10+9
-2=-10+8
-3=-10+7
-4=-10+6
-5=-10+5
-6=-10+4
-7=-10+3
-8=-10+2
-9=-10+1

Let's say we need to subtract 8 from a column but we don't have enough beads.

From the list $-8=-10+2$

10 is the number to unregister , in the next LEFT column (1 lower bead).

2 is the number to register in our column.

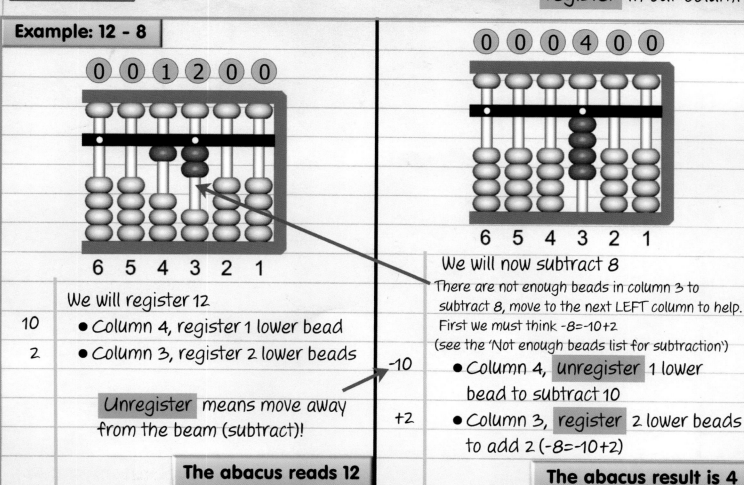

Example: 12 - 8

⓪ ⓪ ① ② ⓪ ⓪

6 5 4 3 2 1

We will register 12

10   ● Column 4, register 1 lower bead

2   ● Column 3, register 2 lower beads

Unregister means move away from the beam (subtract)!

The abacus reads 12

⓪ ⓪ ⓪ ④ ⓪ ⓪

6 5 4 3 2 1

We will now subtract 8

There are not enough beads in column 3 to subtract 8, move to the next LEFT column to help.
First we must think -8=-10+2
(see the 'Not enough beads list for subtraction')

-10   ● Column 4, unregister 1 lower bead to subtract 10

+2   ● Column 3, register 2 lower beads to add 2 (-8=-10+2)

The abacus result is 4

32

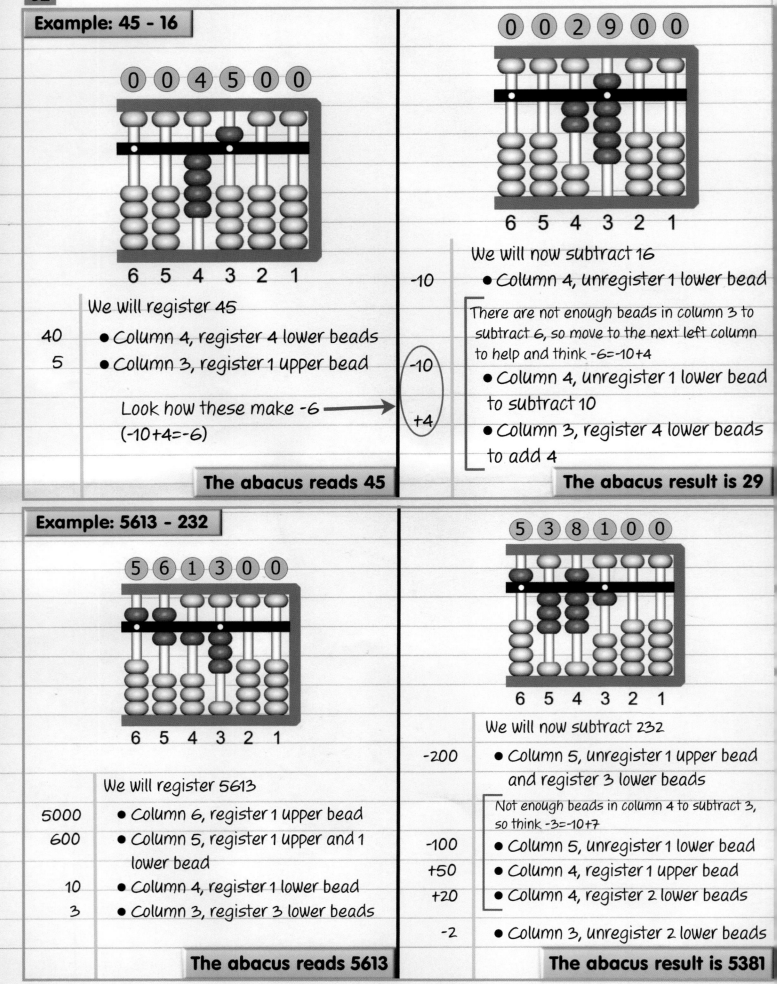

Example: 45 - 16

(0)(0)(4)(5)(0)(0)

6 5 4 3 2 1

We will register 45

40 • Column 4, register 4 lower beads
5 • Column 3, register 1 upper bead

Look how these make -6
(-10+4=-6)

The abacus reads 45

(0)(0)(2)(9)(0)(0)

6 5 4 3 2 1

We will now subtract 16

-10 • Column 4, unregister 1 lower bead

There are not enough beads in column 3 to subtract 6, so move to the next left column to help and think -6=-10+4

-10 • Column 4, unregister 1 lower bead to subtract 10
+4 • Column 3, register 4 lower beads to add 4

The abacus result is 29

Example: 5613 - 232

(5)(6)(1)(3)(0)(0)

6 5 4 3 2 1

We will register 5613

5000 • Column 6, register 1 upper bead
600 • Column 5, register 1 upper and 1 lower bead
10 • Column 4, register 1 lower bead
3 • Column 3, register 3 lower beads

The abacus reads 5613

(5)(3)(8)(1)(0)(0)

6 5 4 3 2 1

We will now subtract 232

-200 • Column 5, unregister 1 upper bead and register 3 lower beads

Not enough beads in column 4 to subtract 3, so think -3=-10+7

-100 • Column 5, unregister 1 lower bead
+50 • Column 4, register 1 upper bead
+20 • Column 4, register 2 lower beads
-2 • Column 3, unregister 2 lower beads

The abacus result is 5381

SKIPPED COLUMNS

Like we did with addition, sometimes with subtraction we have to SKIP a column. When a column doesn't have enough beads left on it to make the subtraction, we move to the next LEFT column to help. Sometimes the next left column also doesn't have enough beads on it, so we SKIP this column and move again to the next left column until you reach a column that has enough beads to use. See below how it works.

- SKIP a column when there are not enough beads to use in that column.

- We will see this symbol ← when we need to SKIP a column (move on to the next left column).

- We will see this symbol → when we need to MOVE BACK a column (move on to the next right column).

- We will REGISTER all beads in any skipped columns
  (with addition we unregistered, here we do the opposite).

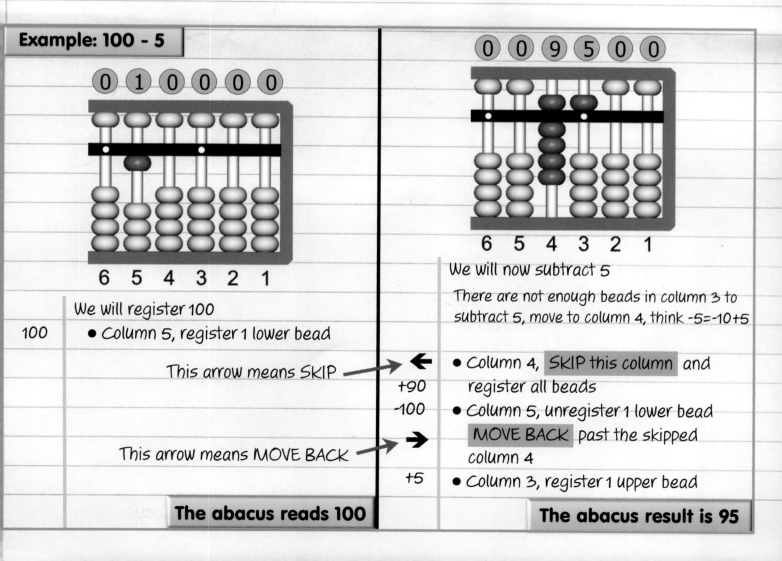

Example: 100 - 5

We will register 100
- Column 5, register 1 lower bead

This arrow means SKIP

This arrow means MOVE BACK

100

The abacus reads 100

We will now subtract 5

There are not enough beads in column 3 to subtract 5, move to column 4, think -5=-10+5

← • Column 4, SKIP this column and register all beads
+90
-100 • Column 5, unregister 1 lower bead
→ MOVE BACK past the skipped column 4
+5 • Column 3, register 1 upper bead

The abacus result is 95

Example: 1000 - 1

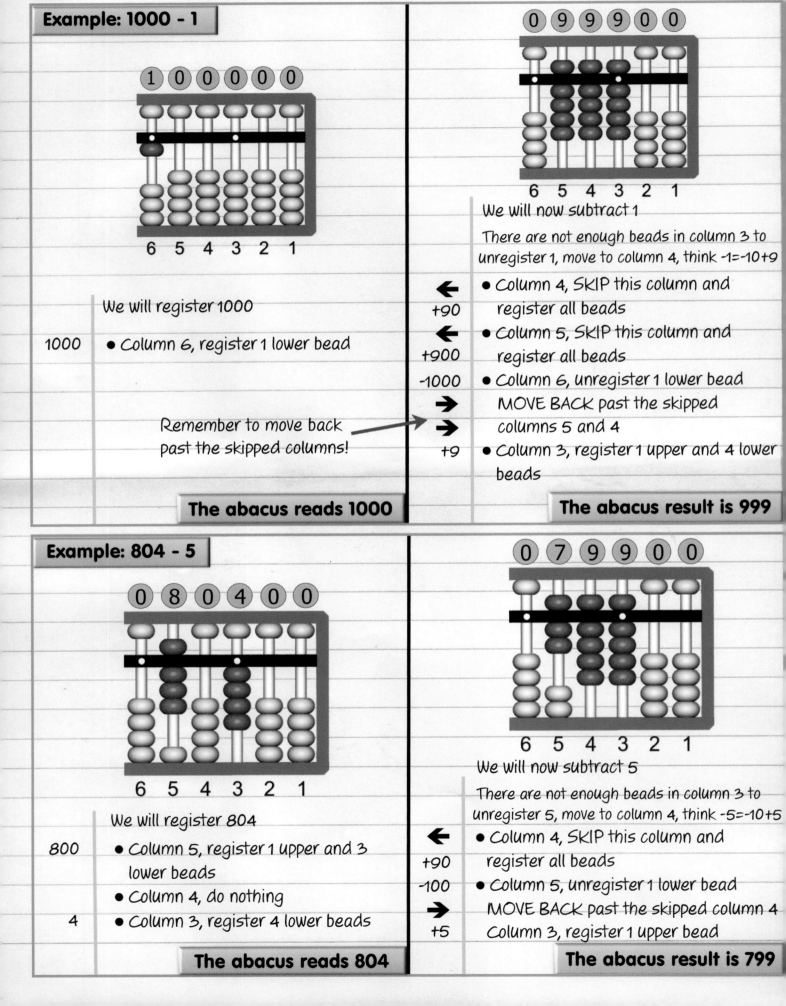

① ⓪ ⓪ ⓪ ⓪ ⓪

6 5 4 3 2 1

We will register 1000

| 1000 | • Column 6, register 1 lower bead |

Remember to move back past the skipped columns!

The abacus reads 1000

⓪ ⑨ ⑨ ⑨ ⓪ ⓪

6 5 4 3 2 1

We will now subtract 1

There are not enough beads in column 3 to unregister 1, move to column 4, think -1=-10+9

← +90	• Column 4, SKIP this column and register all beads
← +900	• Column 5, SKIP this column and register all beads
-1000	• Column 6, unregister 1 lower bead
→ →	MOVE BACK past the skipped columns 5 and 4
+9	• Column 3, register 1 upper and 4 lower beads

The abacus result is 999

Example: 804 - 5

⓪ ⑧ ⓪ ④ ⓪ ⓪

6 5 4 3 2 1

We will register 804

800	• Column 5, register 1 upper and 3 lower beads
	• Column 4, do nothing
4	• Column 3, register 4 lower beads

The abacus reads 804

⓪ ⑦ ⑨ ⑨ ⓪ ⓪

6 5 4 3 2 1

We will now subtract 5

There are not enough beads in column 3 to unregister 5, move to column 4, think -5=-10+5

← +90	• Column 4, SKIP this column and register all beads
-100	• Column 5, unregister 1 lower bead
→	MOVE BACK past the skipped column 4
+5	Column 3, register 1 upper bead

The abacus result is 799

SUBTRACTION OF MORE THAN TWO NUMBERS

When we subtract many numbers on the abacus, just find the difference between the first two numbers, then subtract the next number to get the new difference.

Keep subtracting one number from the difference of the previous numbers until all the numbers have been subtracted.

Example: 998 - 221 - 125

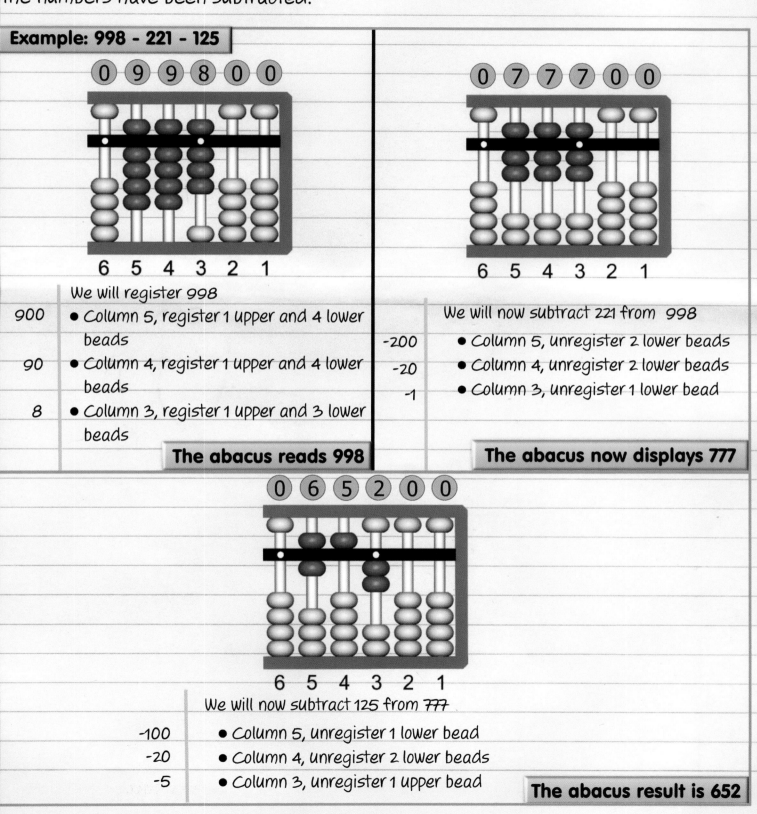

We will register 998

900	• Column 5, register 1 upper and 4 lower beads
90	• Column 4, register 1 upper and 4 lower beads
8	• Column 3, register 1 upper and 3 lower beads

The abacus reads 998

We will now subtract 221 from 998

-200	• Column 5, unregister 2 lower beads
-20	• Column 4, unregister 2 lower beads
-1	• Column 3, unregister 1 lower bead

The abacus now displays 777

We will now subtract 125 from 777

-100	• Column 5, unregister 1 lower bead
-20	• Column 4, unregister 2 lower beads
-5	• Column 3, unregister 1 upper bead

The abacus result is 652

Example: 424662 - 212330 - 1240

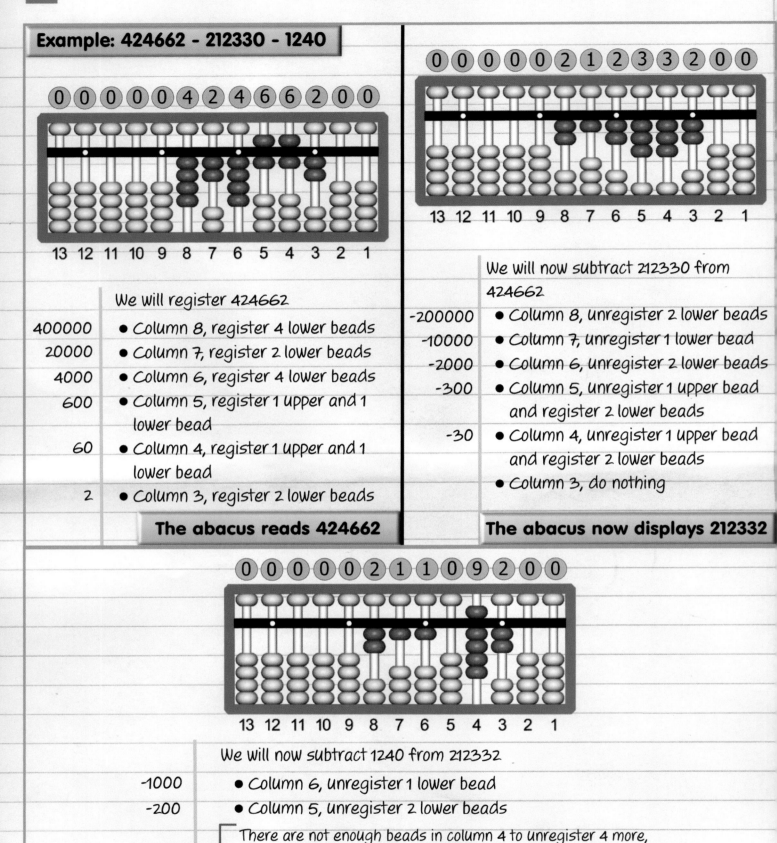

We will register 424662

400000	• Column 8, register 4 lower beads
20000	• Column 7, register 2 lower beads
4000	• Column 6, register 4 lower beads
600	• Column 5, register 1 upper and 1 lower bead
60	• Column 4, register 1 upper and 1 lower bead
2	• Column 3, register 2 lower beads

The abacus reads 424662

We will now subtract 212330 from 424662

-200000	• Column 8, unregister 2 lower beads
-10000	• Column 7, unregister 1 lower bead
-2000	• Column 6, unregister 2 lower beads
-300	• Column 5, unregister 1 upper bead and register 2 lower beads
-30	• Column 4, unregister 1 upper bead and register 2 lower beads
	• Column 3, do nothing

The abacus now displays 212332

We will now subtract 1240 from 212332

-1000	• Column 6, unregister 1 lower bead
-200	• Column 5, unregister 2 lower beads
	There are not enough beads in column 4 to unregister 4 more, move to column 5, think -4=-10+6
-100	• Column 5, unregister 1 lower bead
+60	• Column 4, register 1 upper and 1 lower bead
	• Column 3, do nothing

The abacus result is 211092

SUBTRACTION SUMMARY

- Register your numbers from left to right and in the correct column .

- Sometimes we need to unregister and register in the same column.

- When you don't have enough beads, move to the next LEFT column to help.

- If the next left column also doesn't have enough beads on it, SKIP to the next left column and register all beads in the skipped column (you may need to skip more columns until you reach a column that has enough beads to use, remember to register all beads in any skipped columns).

- When we want to subtract more than two numbers on the abacus, just find the difference between the first two numbers , then subtract the next number to get the new difference .

ADDITION AND SUBTRACTION TOGETHER

Now we will add and subtract numbers in the same calculation, here's how.

We just need to use a combination of adding and subtracting in the same way that we have already learnt to do.

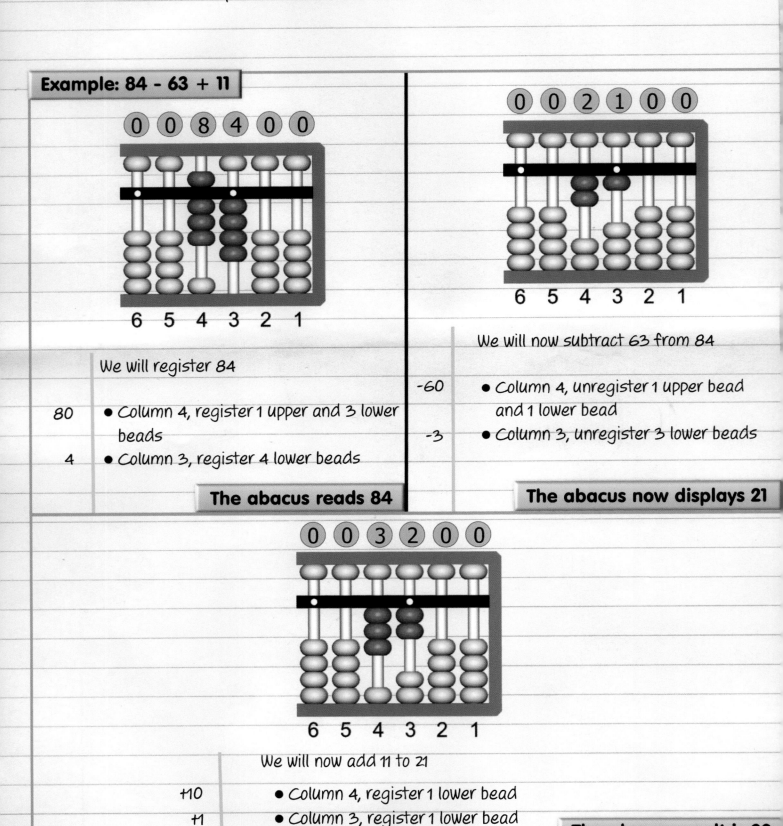

Example: 84 - 63 + 11

0 0 8 4 0 0

6 5 4 3 2 1

We will register 84

| 80 | ● Column 4, register 1 upper and 3 lower beads |
| 4 | ● Column 3, register 4 lower beads |

The abacus reads 84

---

0 0 2 1 0 0

6 5 4 3 2 1

We will now subtract 63 from 84

| -60 | ● Column 4, unregister 1 upper bead and 1 lower bead |
| -3 | ● Column 3, unregister 3 lower beads |

The abacus now displays 21

---

0 0 3 2 0 0

6 5 4 3 2 1

We will now add 11 to 21

| +10 | ● Column 4, register 1 lower bead |
| +1 | ● Column 3, register 1 lower bead |

The abacus result is 32

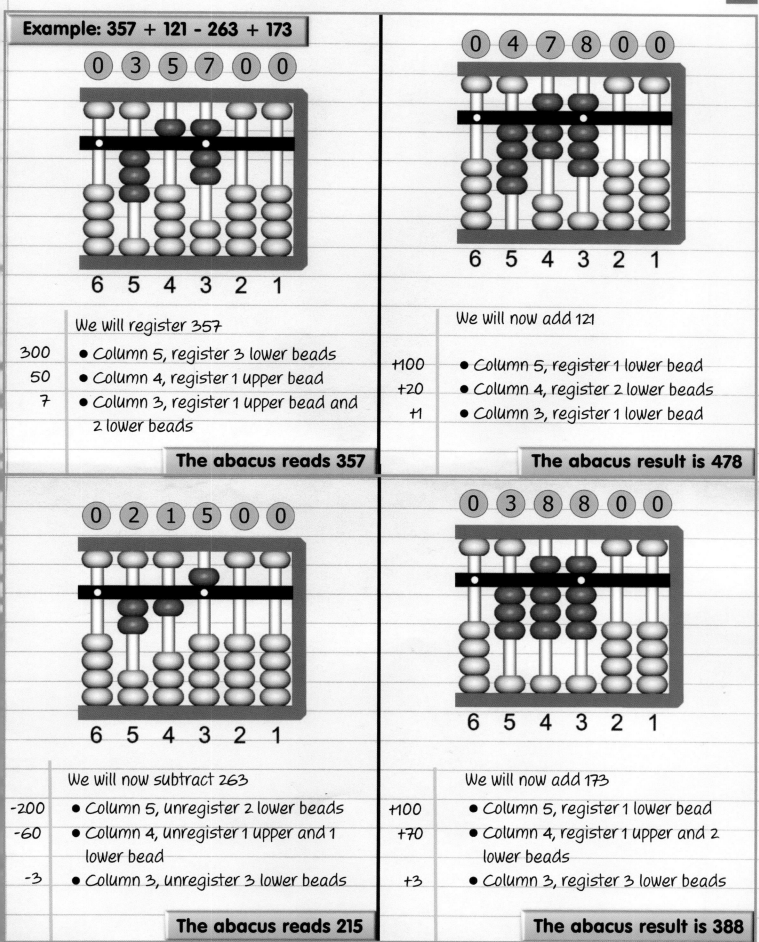

Example: 357 + 121 − 263 + 173

0 3 5 7 0 0

6 5 4 3 2 1

We will register 357

300	• Column 5, register 3 lower beads
50	• Column 4, register 1 upper bead
7	• Column 3, register 1 upper bead and 2 lower beads

The abacus reads 357

0 4 7 8 0 0

6 5 4 3 2 1

We will now add 121

+100	• Column 5, register 1 lower bead
+20	• Column 4, register 2 lower beads
+1	• Column 3, register 1 lower bead

The abacus result is 478

0 2 1 5 0 0

6 5 4 3 2 1

We will now subtract 263

−200	• Column 5, unregister 2 lower beads
−60	• Column 4, unregister 1 upper and 1 lower bead
−3	• Column 3, unregister 3 lower beads

The abacus reads 215

0 3 8 8 0 0

6 5 4 3 2 1

We will now add 173

+100	• Column 5, register 1 lower bead
+70	• Column 4, register 1 upper and 2 lower beads
+3	• Column 3, register 3 lower beads

The abacus result is 388

WORK PAGES

These work pages contain exercises that are meant to be done over-and-over until you are proficient at using the abacus.

- Select a reusable work page of your choice

- Select your column (A, B, C etc..)

WORK PAGE 5

46

A

1	41	20	-18
2	16	22	-12
3	44	32	-15
4	33	16	-12
5	20	22	-20
6	45	41	-12

B

1	15	85	-5
2	23	15	-12
3	60	45	-15
4	26	35	-8
5	14	23	-4
6	44	45	-12

C

1	20	55	-4
2	25	22	-3
3	32	32	-6
4	16	18	-6
5	22	22	-5
6	44	41	-2

- Write down your Page / Column number on the answer sheet
  (see blank answer sheets pages 73 to 83)

- Write down the answers in that column

WRITE THE ANSWERS FOR THE WORK PAGES

73

Page / Column	
5 / B	
1	95
2	26
3	90
4	53
5	33
6	

Page / Column	
1	
2	
3	
4	
5	
6	

Page / Column	
1	
2	
3	
4	
5	
6	

Page / Column	
1	
2	
3	
4	
5	
6	

Page / Column	
1	
2	
3	
4	
5	
6	

- Check your answers with the answer sheet
  (pages 58 to 71)

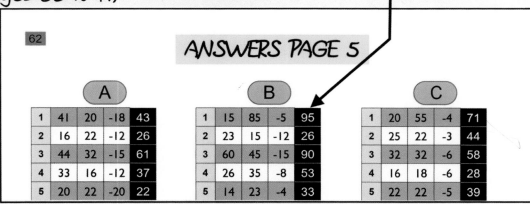

ANSWERS PAGE 5

62

A

1	41	20	-18	43
2	16	22	-12	26
3	44	32	-15	61
4	33	16	-12	37
5	20	22	-20	22

B

1	15	85	-5	95
2	23	15	-12	26
3	60	45	-15	90
4	26	35	-8	53
5	14	23	-4	33

C

1	20	55	-4	71
2	25	22	-3	44
3	32	32	-6	58
4	16	18	-6	28
5	22	22	-5	39

WORK PAGE 1

(Answers to reusable workbook work are on pages 58 to 71)

A				B				C				D		
1	25	20		1	65	55		1	56	55		1	82	55
2	25	22		2	23	22		2	66	25		2	28	25
3	52	33		3	60	32		3	41	32		3	42	32
4	16	12		4	26	12		4	45	16		4	32	22
5	25	22		5	14	12		5	12	10		5	82	22
6	44	20		6	44	41		6	65	44		6	25	18
7	52	41		7	65	52		7	13	12		7	76	42
8	39	25		8	63	37		8	55	36		8	33	32
9	12	9		9	74	15		9	96	12		9	21	20
10	79	75		10	85	75		10	85	79		10	55	25
11	95	82		11	88	82		11	30	19		11	30	19
12	32	14		12	36	18		12	30	32		12	36	32
13	65	42		13	54	42		13	25	24		13	65	24
14	12	32		14	26	32		14	30	12		14	12	12
15	72	45		15	85	20		15	65	72		15	72	72
16	85	65		16	55	25		16	54	32		16	85	32
17	25	22		17	92	6		17	47	6		17	25	6
18	10	8		18	75	12		18	60	12		18	39	36
19	11	5		19	76	11		19	13	11		19	55	54
20	51	21		20	65	55		20	56	55		20	51	26
21	64	64		21	66	64		21	96	64		21	64	23
22	48	45		22	49	22		22	55	45		22	32	30
23	85	65		23	85	66		23	44	23		23	92	5
24	48	18		24	91	18		24	22	18		24	75	6
25	12	8		25	21	10		25	45	10		25	76	20
26	8	3		26	26	5		26	85	3		26	85	65
27	21	1		27	26	6		27	58	9		27	58	8
28	99	98		28	52	23		28	99	12		28	99	14
29	47	12		29	58	47		29	90	47		29	90	47
30	76	3		30	56	26		30	88	3		30	77	11

WORK PAGE 2

A

1	21	55
2	31	25
3	41	32
4	18	22
5	25	36
6	45	18
7	52	66
8	32	32
9	12	12
10	77	25
11	95	21
12	31	32
13	65	25
14	12	13
15	72	77
16	85	32
17	85	6
18	10	66
19	11	54
20	52	26
21	64	23
22	48	32
23	85	9
24	78	6
25	12	22
26	8	55
27	21	8
28	99	15
29	88	47
30	76	12

B

1	52	55
2	63	22
3	41	32
4	42	12
5	96	12
6	85	41
7	30	55
8	30	37
9	66	15
10	30	75
11	65	88
12	54	18
13	55	42
14	60	32
15	13	22
16	56	25
17	99	6
18	60	12
19	13	12
20	56	56
21	96	77
22	55	44
23	44	66
24	33	18
25	45	10
26	85	8
27	58	6
28	99	25
29	25	47
30	26	26

C

1	56	65
2	66	23
3	41	60
4	45	26
5	12	14
6	65	44
7	13	65
8	55	63
9	96	74
10	85	85
11	30	88
12	30	36
13	25	54
14	30	26
15	65	85
16	54	55
17	47	92
18	60	75
19	13	76
20	56	65
21	96	66
22	55	49
23	44	85
24	22	91
25	45	21
26	85	26
27	58	26
28	99	52
29	90	58
30	76	56

D

1	33	56
2	30	66
3	25	45
4	30	45
5	65	12
6	66	65
7	47	13
8	60	35
9	13	96
10	55	88
11	30	30
12	36	30
13	64	25
14	12	39
15	72	45
16	85	54
17	52	47
18	39	60
19	55	34
20	51	56
21	58	96
22	32	55
23	88	45
24	75	22
25	76	25
26	85	9
27	25	58
28	99	5
29	45	90
30	88	76

WORK PAGE 3

A		
1	25	-20
2	25	-13
3	52	-30
4	85	-41
5	25	-9
6	66	-54
7	52	-13
8	39	-20
9	74	-20
10	79	-15
11	95	-65
12	32	-32
13	65	-7
14	12	-8
15	72	-45
16	85	-65
17	25	-22
18	10	-8
19	11	-5
20	51	-40
21	64	-33
22	48	-3
23	85	-2
24	48	-7
25	12	-5
26	88	-41
27	21	-14
28	99	-2
29	47	-12
30	76	-3

B		
1	42	-3
2	32	-6
3	45	-8
4	65	-4
5	22	-3
6	37	-15
7	64	-32
8	51	-41
9	64	-50
10	45	-6
11	65	-3
12	64	-23
13	12	-10
14	94	-4
15	55	-47
16	55	-30
17	92	-53
18	75	-23
19	76	-55
20	65	-5
21	85	-70
22	49	-19
23	85	-14
24	91	-5
25	21	-6
26	26	-1
27	26	-12
28	52	-7
29	58	-47
30	56	-26

C		
1	66	-33
2	63	-35
3	39	-12
4	45	-8
5	22	-12
6	65	-5
7	44	-41
8	55	-8
9	96	-15
10	85	-9
11	30	-4
12	30	-12
13	25	-16
14	88	-74
15	65	-33
16	98	-66
17	47	-6
18	60	-13
19	13	-2
20	56	-51
21	96	-12
22	55	-44
23	44	-33
24	22	-3
25	45	-2
26	85	-9
27	58	-5
28	99	-41
29	90	-14
30	76	-2

D		
1	55	-10
2	20	-8
3	40	-14
4	32	-13
5	82	-9
6	25	-18
7	76	-15
8	33	-14
9	66	-44
10	55	-12
11	30	-16
12	36	-16
13	65	-45
14	36	-12
15	72	-33
16	85	-25
17	25	-12
18	39	-10
19	55	-24
20	51	-8
21	64	-12
22	32	-5
23	92	-41
24	75	-8
25	76	-15
26	85	-9
27	58	-4
28	99	-12
29	90	-23
30	77	-6

A

#		
1	44	-10
2	35	-8
3	45	-14
4	65	-13
5	25	-9
6	37	-4
7	64	-15
8	44	-14
9	64	-25
10	45	-12
11	66	-16
12	32	-16
13	65	-32
14	12	-12
15	72	-33
16	88	-25
17	25	-12
18	10	-8
19	55	-25
20	51	-8
21	66	-12
22	48	-8
23	85	-41
24	44	-8
25	12	-11
26	88	-9
27	32	-4
28	99	-14
29	47	-23
30	77	-6

B

#		
1	55	-33
2	55	-35
3	77	-12
4	85	-8
5	30	-14
6	44	-5
7	99	-88
8	88	-8
9	65	-15
10	78	-9
11	47	-7
12	64	-12
13	85	-16
14	94	-74
15	55	-21
16	87	-66
17	92	-6
18	75	-13
19	76	-7
20	77	-51
21	85	-15
22	21	-12
23	85	-33
24	91	-3
25	21	-2
26	55	-9
27	26	-8
28	52	-41
29	78	-14
30	56	-4

C

#		
1	66	-3
2	85	-13
3	25	-7
4	64	-4
5	55	-18
6	51	-15
7	77	-32
8	32	-21
9	92	-50
10	37	-6
11	76	-3
12	85	-23
13	58	-10
14	99	-9
15	90	-47
16	88	-30
17	96	-53
18	60	-5
19	65	-55
20	56	-5
21	77	-70
22	55	-35
23	44	-14
24	55	-5
25	45	-7
26	85	-15
27	75	-64
28	80	-7
29	90	-80
30	79	-26

D

#		
1	67	-20
2	55	-26
3	92	-30
4	75	-43
5	84	-9
6	65	-32
7	97	-13
8	33	-25
9	94	-20
10	55	-27
11	99	-65
12	36	-14
13	66	-8
14	36	-9
15	84	-45
16	85	-36
17	25	-22
18	39	-7
19	49	-5
20	51	-31
21	64	-33
22	91	-3
23	92	-74
24	83	-7
25	76	-4
26	73	-42
27	58	-14
28	99	-15
29	90	-75
30	82	-43

WORK PAGE 5

A

1	41	20	-18
2	16	22	-12
3	44	32	-15
4	33	16	-12
5	20	22	-20
6	45	41	-12
7	72	52	-32
8	62	36	-41
9	80	12	-8
10	14	75	-55
11	23	82	-12
12	36	14	-19
13	20	42	-19
14	66	32	-15
15	45	45	-62
16	55	65	-32
17	12	22	-5
18	41	10	-8
19	25	11	-6
20	55	51	-2
21	99	64	-45
22	32	45	-15
23	15	65	-15
24	13	18	-6
25	47	12	-6
26	88	3	-12
27	66	1	-15
28	82	98	-75
29	92	47	-33
30	24	76	-66

B

1	15	85	-5
2	23	15	-12
3	60	45	-15
4	26	35	-8
5	14	23	-4
6	44	45	-12
7	55	74	-12
8	63	62	-25
9	74	82	-9
10	46	14	-54
11	14	38	-13
12	36	33	-20
13	14	20	-20
14	26	65	-16
15	85	45	-65
16	55	54	-32
17	92	12	-7
18	75	12	-8
19	76	13	-3
20	65	55	-2
21	32	96	-45
22	26	32	-14
23	85	12	-33
24	91	13	-4
25	21	47	-2
26	26	85	-9
27	26	66	-5
28	4	82	-42
29	18	91	-14
30	9	23	-2

C

1	20	55	-4
2	25	22	-3
3	32	32	-6
4	16	18	-6
5	22	22	-5
6	44	41	-2
7	52	52	-6
8	36	37	-10
9	12	12	-9
10	79	75	-22
11	19	82	-2
12	32	18	-12
13	24	42	-6
14	12	32	-4
15	72	20	-3
16	32	65	-10
17	64	6	-24
18	10	11	-13
19	11	11	-2
20	51	55	-51
21	64	64	-12
22	48	45	-44
23	65	66	-33
24	18	18	-3
25	12	10	-2
26	32	3	-9
27	35	1	-5
28	98	99	-41
29	47	47	-14
30	76	76	-2

WORK PAGE 6

A

#			
1	52	74	-18
2	16	52	-12
3	33	82	-15
4	33	16	-12
5	20	38	-13
6	62	52	-12
7	72	21	-32
8	62	65	-32
9	80	40	-8
10	16	54	-55
11	23	12	-9
12	36	8	-19
13	44	13	-19
14	66	52	-15
15	45	96	-52
16	55	12	-32
17	44	8	-5
18	41	6	-8
19	25	47	-6
20	55	85	-2
21	99	66	-33
22	32	20	-15
23	15	91	-15
24	13	16	-6
25	47	12	-4
26	99	3	-12
27	66	3	-15
28	82	98	-13
29	92	47	-33
30	24	20	-23

B

#			
1	22	85	-5
2	23	82	-12
3	60	12	-16
4	16	12	-8
5	22	47	-4
6	20	85	-9
7	52	16	-8
8	36	88	-25
9	12	91	-9
10	79	23	-44
11	19	17	-13
12	32	9	-20
13	24	9	-18
14	12	28	-12
15	16	47	-15
16	32	76	-5
17	64	22	-20
18	10	55	-12
19	80	13	-32
20	14	30	-41
21	23	96	-6
22	36	32	-55
23	20	11	-12
24	66	11	-4
25	45	47	-16
26	55	55	-9
27	77	66	-25
28	4	82	-42
29	18	36	-14
30	9	39	-2

C

#			
1	10	55	-4
2	10	22	-3
3	32	36	-6
4	16	18	-3
5	22	22	-20
6	44	20	-12
7	64	50	-15
8	20	37	-12
9	11	12	-20
10	51	68	-12
11	64	55	-32
12	95	18	-41
13	85	42	-8
14	91	13	-55
15	21	5	-3
16	26	60	-10
17	28	6	-24
18	4	20	-13
19	11	41	-2
20	60	55	-51
21	64	64	-12
22	48	8	-44
23	65	76	-33
24	18	55	-3
25	99	10	-2
26	32	6	-9
27	35	12	-5
28	98	41	-41
29	47	20	-14
30	76	88	-2

WORK PAGE 7

A

#			
1	88	74	-3
2	15	52	-3
3	48	82	-6
4	35	16	-7
5	25	38	-5
6	45	55	-2
7	75	21	-6
8	62	25	-12
9	82	40	-9
10	14	54	-22
11	39	12	-15
12	33	8	-12
13	20	15	-6
14	65	52	-10
15	47	96	-3
16	54	12	-10
17	18	9	-24
18	12	6	-13
19	46	47	-2
20	55	85	-51
21	96	67	-12
22	32	20	-44
23	12	91	-33
24	13	16	-14
25	88	12	-2
26	85	3	-20
27	66	12	-5
28	82	98	-44
29	85	47	-14
30	23	26	-12

B

#			
1	66	85	-15
2	85	88	-12
3	25	12	-15
4	65	12	-18
5	55	47	-20
6	51	95	-13
7	85	16	-20
8	32	88	-41
9	92	99	-7
10	37	23	-45
11	88	17	-13
12	85	15	-15
13	58	9	-22
14	99	36	-15
15	90	47	-65
16	99	76	-13
17	96	22	-7
18	60	56	-6
19	65	15	-9
20	56	30	-4
21	87	96	-26
22	55	32	-12
23	44	15	-15
24	55	11	-7
25	46	47	-4
26	85	60	-12
27	85	66	-16
28	80	82	-44
29	90	37	-33
30	79	42	-55

C

#			
1	55	55	-6
2	30	25	-8
3	25	36	-7
4	33	18	-4
5	65	40	-12
6	66	20	-12
7	47	55	-22
8	60	37	-42
9	45	12	-30
10	55	72	-6
11	30	55	-3
12	36	18	-33
13	64	42	-10
14	24	13	-6
15	72	5	-35
16	85	60	-30
17	55	6	-42
18	39	25	-23
19	55	41	-55
20	51	55	-42
21	65	64	-12
22	32	28	-5
23	88	76	-72
24	85	55	-18
25	76	12	-12
26	85	6	-15
27	25	12	-9
28	88	52	-9
29	45	20	-7
30	75	88	-8

WORK PAGE 8

A

1	10	20	-15	-5
2	30	22	-12	-3
3	40	32	-15	-2
4	5	16	-14	-4
5	10	22	-20	-5
6	30	41	-13	-1
7	10	52	-30	-5
8	30	36	-41	-10
9	10	12	-9	-9
10	30	78	-54	-20
11	10	19	-13	-2
12	30	32	-20	-12
13	10	24	-20	-5
14	30	6	-15	-4
15	10	72	-65	-3
16	30	32	-32	-17
17	42	6	-7	-24
18	30	10	-8	-13
19	10	11	-3	-4
20	30	51	-2	-51
21	10	64	-45	-12
22	30	45	-12	-40
23	10	65	-15	-33
24	30	18	-6	-3
25	10	12	-4	-2
26	30	3	-12	-7
27	47	1	-15	-5
28	30	98	-74	-41
29	10	47	-33	-14
30	30	76	-66	-2

B

1	15	-3	63	-5
2	22	-6	15	-3
3	60	-8	42	-2
4	26	-4	35	-4
5	12	-3	27	-5
6	44	-12	45	-1
7	52	-32	65	-5
8	63	-41	75	-10
9	74	-50	82	-9
10	21	-5	14	-20
11	12	-3	38	-2
12	36	-23	32	-12
13	14	-10	20	-5
14	25	-7	65	-4
15	85	-47	41	-3
16	52	-30	54	-17
17	92	-53	12	-24
18	74	-23	9	-13
19	76	-55	13	-4
20	65	-42	54	-51
21	32	-14	96	-12
22	25	-6	32	-40
23	85	-70	9	-6
24	91	-19	2	-3
25	21	-14	47	-2
26	25	-3	85	-7
27	26	-6	61	-5
28	2	-1	82	-41
29	18	-14	90	-14
30	9	-7	23	-2

C

1	20	20	-10	-5
2	65	22	-8	-12
3	41	32	-13	-15
4	54	16	-14	-6
5	12	22	-7	-4
6	9	41	-8	-12
7	13	52	-3	-15
8	54	36	-2	-74
9	96	12	-45	-9
10	30	75	-12	-54
11	10	82	-15	-13
12	30	14	-6	-20
13	10	42	-4	-20
14	30	32	-12	-15
15	65	20	-32	-45
16	54	65	-35	-32
17	45	6	-12	-7
18	60	10	-7	-8
19	13	11	-12	-3
20	54	51	-5	-2
21	96	64	-41	-45
22	32	45	-8	-12
23	9	65	-15	-33
24	2	18	-6	-3
25	47	12	-4	-2
26	85	3	-12	-7
27	61	1	-15	-5
28	82	98	-74	-41
29	90	47	-33	-14
30	23	76	-66	-2

WORK PAGE 9

A

#				
1	85	20	-15	-5
2	15	22	-12	-3
3	45	32	-15	-2
4	35	16	-14	-4
5	23	22	-20	-5
6	45	41	-13	-1
7	74	52	-30	-5
8	62	36	-41	-10
9	82	12	-9	-9
10	14	75	-54	-20
11	38	82	-13	-2
12	33	14	-20	-12
13	20	42	-20	-5
14	65	32	-15	-4
15	45	45	-65	-3
16	54	65	-32	-17
17	12	22	-7	-24
18	12	10	-8	-13
19	13	11	-3	-4
20	55	51	-2	-51
21	96	64	-45	-12
22	32	45	-12	-40
23	12	65	-15	-33
24	13	18	-6	-3
25	47	12	-4	-2
26	85	3	-12	-7
27	66	1	-15	-5
28	82	98	-74	-41
29	91	47	-33	-14
30	23	76	-66	-2

B

#				
1	15	-3	20	-5
2	23	-6	25	-12
3	60	-8	32	-15
4	26	-4	16	-8
5	14	-3	22	-4
6	44	-15	44	-12
7	55	-32	52	-12
8	63	-41	36	-25
9	74	-50	12	-9
10	21	-6	79	-54
11	14	-3	19	-13
12	36	-23	32	-20
13	14	-10	24	-20
14	26	-4	12	-16
15	85	-47	72	-65
16	55	-30	32	-32
17	92	-53	6	-7
18	75	-23	10	-8
19	76	-55	11	-3
20	65	-42	51	-2
21	32	-14	64	-45
22	26	-5	48	-14
23	85	-70	65	-33
24	91	-19	18	-4
25	21	-14	12	-2
26	26	-5	3	-9
27	26	-6	1	-5
28	4	-1	98	-42
29	18	-12	47	-14
30	9	-7	76	-2

C

#				
1	21	55	-10	-4
2	66	22	-8	-3
3	41	32	-14	-6
4	45	18	-14	-6
5	12	22	-9	-5
6	9	41	-8	-2
7	13	52	-3	-6
8	55	37	-2	-10
9	96	12	-44	-9
10	31	75	-12	-22
11	10	82	-16	-2
12	30	18	-6	-12
13	12	42	-4	-6
14	30	32	-12	-4
15	65	20	-33	-3
16	54	65	-35	-10
17	47	6	-12	-24
18	60	11	-8	-13
19	13	11	-12	-2
20	54	55	-5	-51
21	96	64	-41	-12
22	32	45	-8	-44
23	11	66	-15	-33
24	2	18	-9	-3
25	45	10	-4	-2
26	85	3	-12	-9
27	58	1	-16	-5
28	82	99	-74	-41
29	90	47	-33	-14
30	25	76	-66	-2

WORK PAGE 10

A

#				
1	30	20	-15	-4
2	65	22	-15	-3
3	41	25	-15	-5
4	54	16	-23	-4
5	14	22	-20	-5
6	9	44	-14	-1
7	16	52	-30	-7
8	54	74	-39	-10
9	99	12	-9	-9
10	30	77	-54	-20
11	12	82	-13	-12
12	30	15	-21	-12
13	10	42	-20	-5
14	66	32	-12	-4
15	65	52	-65	-3
16	54	65	-25	-17
17	52	22	-7	-24
18	60	12	-9	-13
19	13	11	-3	-4
20	54	44	-6	-51
21	78	64	-45	-15
22	60	32	-15	-40
23	9	65	-15	-33
24	2	25	-6	-3
25	52	12	-4	-2
26	85	3	-12	-11
27	61	1	-15	-5
28	82	88	-74	-41
29	99	47	-33	-16
30	23	80	-66	-3

B

#				
1	40	-5	40	-5
2	22	-15	25	-10
3	32	-32	25	-12
4	16	-8	16	-8
5	41	-5	22	-4
6	55	-25	26	-12
7	52	-40	53	-14
8	36	-23	33	-25
9	12	-11	15	-9
10	75	-45	70	-54
11	87	-11	20	-13
12	63	-27	33	-20
13	42	-20	13	-20
14	32	-18	54	-16
15	84	-74	72	-66
16	65	-25	60	-32
17	16	-8	20	-7
18	10	-8	15	-8
19	11	-5	12	-3
20	66	-2	12	-2
21	64	-20	64	-45
22	45	-14	24	-14
23	65	-25	66	-33
24	18	-9	18	-4
25	24	-2	21	-2
26	35	-10	9	-9
27	36	-6	3	-5
28	98	-3	98	-42
29	47	-14	54	-14
30	45	-7	25	-2

C

#				
1	36	55	-4	-15
2	25	32	-3	-20
3	56	32	-6	-15
4	16	19	-6	-13
5	62	22	-20	-20
6	25	42	-12	-13
7	52	52	-17	-30
8	33	52	-12	-32
9	12	35	-20	-9
10	74	75	-10	-54
11	19	88	-32	-13
12	52	18	-42	-20
13	32	42	-8	-20
14	54	32	-45	-15
15	72	20	-3	-12
16	60	55	-10	-32
17	18	18	-24	-7
18	15	11	-13	-8
19	12	11	-2	-3
20	12	65	-51	-2
21	64	25	-12	-45
22	48	45	-40	-12
23	66	66	-33	-9
24	18	27	-3	-6
25	88	10	-2	-4
26	18	3	-5	-12
27	20	8	-5	-15
28	98	99	-32	-74
29	36	47	-14	-51
30	35	25	-2	-33

WORK PAGE 11

A

1	88	20	-12	-2	-5
2	17	22	-12	-4	-3
3	44	33	-15	-4	-2
4	30	16	-16	-4	-4
5	30	22	-20	-9	-5
6	44	44	-13	-1	-1
7	77	52	-20	-5	-5
8	63	26	-41	-12	-10
9	82	12	-9	-12	-9
10	15	78	-45	-22	-20
11	33	82	-13	-3	-2
12	35	15	-15	-12	-12
13	25	42	-20	-5	-5
14	66	33	-15	-4	-4
15	44	45	-65	-6	-3
16	50	66	-16	-17	-17
17	40	22	-9	-24	-24
18	30	12	-9	-13	-13
19	78	11	-9	-4	-4
20	90	51	-12	-25	-51
21	95	64	-25	-12	-12
22	67	47	-12	-30	-40
23	46	65	-15	-33	-33
24	28	78	-7	-9	-3
25	82	12	-4	-2	-2
26	65	36	-12	-17	-7
27	46	6	-16	-5	-5
28	95	14	-44	-21	-41
29	52	55	-33	-33	-14
30	32	85	-55	-6	-2

B

1	15	-6	40	-5	14
2	22	-6	25	-12	25
3	60	-7	33	-32	30
4	26	-4	16	-8	16
5	15	-12	22	-4	36
6	44	-15	25	-12	44
7	56	-22	52	-40	52
8	63	-41	33	-25	36
9	74	-30	12	-9	12
10	22	-6	70	-54	79
11	14	-3	19	-11	19
12	35	-23	33	-20	33
13	15	-10	24	-20	24
14	26	-6	54	-16	12
15	88	-47	72	-70	62
16	55	-30	60	-35	33
17	99	-42	6	-8	6
18	75	-23	12	-8	14
19	85	-50	12	-3	11
20	65	-42	12	-2	52
21	32	-12	64	-60	66
22	13	-5	48	-14	44
23	85	-70	66	-30	65
24	91	-18	18	-4	44
25	20	-12	22	-2	12
26	26	-9	9	-10	8
27	36	-9	3	-5	1
28	14	-9	98	-25	44
29	18	-6	25	-14	25
30	10	-8	25	-6	12

WORK PAGE 12

A

1	20	20	-4	-2	20
2	65	22	-3	-4	22
3	41	33	-6	-4	32
4	54	16	-3	-4	16
5	12	22	-20	-9	22
6	9	44	-12	-1	41
7	13	52	-15	-5	52
8	54	26	-12	-12	36
9	96	12	-20	-12	12
10	30	78	-12	-22	75
11	10	82	-32	-3	82
12	30	33	-41	-12	14
13	10	42	-8	-5	42
14	30	33	-55	-4	32
15	65	45	-3	-6	20
16	54	66	-10	-17	65
17	45	22	-24	-24	6
18	60	12	-13	-13	10
19	13	11	-2	-4	11
20	54	51	-51	-25	51
21	96	64	-12	-12	64
22	60	47	-44	-30	45
23	9	65	-33	-33	65
24	2	78	-3	-9	18
25	47	12	-2	-2	12
26	85	36	-9	-17	3
27	61	6	-5	-5	1
28	82	14	-41	-21	98
29	90	55	-14	-33	47
30	23	85	-2	-6	76

B

1	41	-15	40	-5	14
2	16	-12	25	-12	25
3	44	-15	33	-32	30
4	33	-14	16	-8	16
5	20	-20	22	-4	36
6	45	-13	25	-26	44
7	72	-30	52	-40	52
8	62	-41	33	-25	36
9	80	-9	12	-9	12
10	64	-54	70	-54	79
11	23	-13	19	-11	19
12	36	-20	33	-38	33
13	36	-20	22	-20	24
14	66	-15	54	-16	12
15	45	-3	72	-70	62
16	55	-32	60	-35	33
17	85	-7	6	-8	6
18	41	-8	12	-8	14
19	25	-3	12	-3	11
20	55	-2	12	-2	52
21	99	-45	64	-60	66
22	32	-12	48	-14	44
23	16	-15	66	-30	65
24	13	-6	18	-4	44
25	47	-4	22	-2	12
26	88	-12	9	-10	8
27	66	-15	3	-5	1
28	82	-74	98	-25	44
29	92	-33	25	-14	25
30	99	-66	25	-6	12

WORK PAGE 13

A

1	102	242	-3	-200	20
2	22	12	-6	-4	22
3	123	62	-8	-4	32
4	18	121	-100	-4	16
5	22	6	-3	-9	22
6	441	14	-15	-210	41
7	225	11	-32	-5	52
8	550	111	-41	-221	36
9	12	66	-50	-12	12
10	321	44	-6	-22	75
11	82	65	-3	-3	82
12	18	623	-23	-12	14
13	42	42	-10	-22	42
14	251	33	-4	-4	32
15	20	45	-47	-6	20
16	65	412	-30	-17	65
17	333	22	-53	-120	6
18	412	12	-23	-13	10
19	11	223	-55	-4	11
20	550	51	-400	-25	51
21	213	821	-70	-300	64
22	45	90	-19	-60	45
23	166	65	-121	-33	65
24	18	78	-5	-9	18
25	362	12	-6	-2	12
26	875	36	-418	-17	3
27	365	6	-12	-5	1
28	999	14	-555	-21	98
29	400	55	-47	-33	47
30	76	288	-200	-6	76

B

1	300	-15	40	-10	14
2	22	-12	520	-8	25
3	152	-140	33	-14	30
4	36	-14	160	-13	16
5	220	-20	22	-9	36
6	75	-13	125	-18	44
7	882	-255	52	-15	52
8	140	-41	33	-44	36
9	425	-250	12	-44	12
10	632	-155	70	-120	79
11	120	-13	19	-16	19
12	65	-20	33	-16	33
13	600	-200	22	-310	24
14	100	-15	54	-12	12
15	110	-35	72	-33	62
16	665	-32	60	-150	33
17	452	-200	6	-120	6
18	210	-8	12	-10	14
19	110	-3	130	-24	11
20	230	-25	12	-8	52
21	64	-45	64	-12	66
22	125	-12	233	-155	44
23	610	-250	66	-200	65
24	180	-80	18	-8	44
25	120	-4	22	-15	12
26	125	-25	9	-9	8
27	225	-106	3	-40	1
28	99	-25	188	-12	44
29	850	-450	25	-23	25
30	320	-120	25	-6	12

WORK PAGE 14

A

1	120	242	-3	-12	20
2	222	100	-120	-4	22
3	133	62	-8	-25	32
4	16	121	-100	-4	16
5	250	6	-120	-9	22
6	44	320	-15	-120	41
7	520	225	-365	-5	52
8	26	111	-41	-12	36
9	120	66	-50	-120	12
10	780	44	-400	-22	75
11	82	665	-3	-3	82
12	330	623	-23	-12	14
13	42	250	-25	-5	42
14	124	33	-4	-24	32
15	450	45	-200	-6	20
16	66	412	-30	-17	65
17	222	22	-30	-24	6
18	12	444	-23	-44	10
19	110	223	-55	-4	11
20	51	515	-400	-25	51
21	132	821	-70	-520	64
22	165	90	-19	-30	45
23	65	241	-121	-14	65
24	136	78	-5	-9	18
25	12	444	-6	-2	12
26	950	555	-418	-600	3
27	963	6	-12	-540	1
28	652	100	-555	-21	98
29	55	200	-47	-33	47
30	425	288	-200	-6	76

B

1	240	-6	120	-10	14
2	125	-6	420	-8	25
3	130	-7	55	-14	30
4	160	-4	160	-13	16
5	360	-12	140	-145	36
6	440	-15	125	-250	44
7	225	-200	52	-15	52
8	635	-410	33	-44	36
9	127	-30	12	-44	12
10	790	-650	70	-120	79
11	250	-54	19	-16	19
12	330	-23	33	-16	33
13	240	-110	200	-310	24
14	120	-60	54	-12	12
15	665	-470	72	-33	62
16	352	-300	500	-150	33
17	362	-320	600	-120	6
18	145	-23	12	-10	14
19	110	-50	130	-24	11
20	520	-420	12	-8	52
21	660	-500	64	-12	66
22	415	-350	233	-155	44
23	254	-145	66	-20	65
24	400	-200	18	-8	44
25	120	-25	22	-15	12
26	880	-741	9	-9	8
27	199	-99	3	-40	1
28	420	-145	188	-12	44
29	250	-125	25	-23	25
30	150	-50	25	-6	12

ANSWERS

58

ANSWERS PAGE 1

A

#			
1	25	20	45
2	25	22	47
3	52	33	85
4	16	12	28
5	25	22	47
6	44	20	64
7	52	41	93
8	39	25	64
9	12	9	21
10	79	75	154
11	95	82	177
12	32	14	46
13	65	42	107
14	12	32	44
15	72	45	117
16	85	65	150
17	25	22	47
18	10	8	18
19	11	5	16
20	51	21	72
21	64	64	128
22	48	45	93
23	85	65	150
24	48	18	66
25	12	8	20
26	8	3	11
27	21	1	22
28	99	98	197
29	47	12	59
30	76	3	79

B

#			
1	65	55	120
2	23	22	45
3	60	32	92
4	26	12	38
5	14	12	26
6	44	41	85
7	65	52	117
8	63	37	100
9	74	15	89
10	85	75	160
11	88	82	170
12	36	18	54
13	54	42	96
14	26	32	58
15	85	20	105
16	55	25	80
17	92	6	98
18	75	12	87
19	76	11	87
20	65	55	120
21	66	64	130
22	49	22	71
23	85	66	151
24	91	18	109
25	21	10	31
26	26	5	31
27	26	6	32
28	52	23	75
29	58	47	105
30	56	26	82

C

#			
1	56	55	111
2	66	25	91
3	41	32	73
4	45	16	61
5	12	10	22
6	65	44	109
7	13	12	25
8	55	36	91
9	96	12	108
10	85	79	164
11	30	19	49
12	30	32	62
13	25	24	49
14	30	12	42
15	65	72	137
16	54	32	86
17	47	6	53
18	60	12	72
19	13	11	24
20	56	55	111
21	96	64	160
22	55	45	100
23	44	23	67
24	22	18	40
25	45	10	55
26	85	3	88
27	58	9	67
28	99	12	111
29	90	47	137
30	88	3	91

D

#			
1	82	55	137
2	28	25	53
3	42	32	74
4	32	22	54
5	82	22	104
6	25	18	43
7	76	42	118
8	33	32	65
9	21	20	41
10	55	25	80
11	30	19	49
12	36	32	68
13	65	24	89
14	12	12	24
15	72	72	144
16	85	32	117
17	25	6	31
18	39	36	75
19	55	54	109
20	51	26	77
21	64	23	87
22	32	30	62
23	92	5	97
24	75	6	81
25	76	20	96
26	85	65	150
27	58	8	66
28	99	14	113
29	90	47	137
30	77	11	88

A

#			
1	21	55	76
2	31	25	56
3	41	32	73
4	18	22	40
5	25	36	61
6	45	18	63
7	52	66	118
8	32	32	64
9	12	12	24
10	77	25	102
11	95	21	116
12	31	32	63
13	65	25	90
14	12	13	25
15	72	77	149
16	85	32	117
17	85	6	91
18	10	66	76
19	11	54	65
20	52	26	78
21	64	23	87
22	48	32	80
23	85	9	94
24	78	6	84
25	12	22	34
26	8	55	63
27	21	8	29
28	99	15	114
29	88	47	135
30	76	12	88

B

#			
1	52	55	107
2	63	22	85
3	41	32	73
4	42	12	54
5	96	12	108
6	85	41	126
7	30	55	85
8	30	37	67
9	66	15	81
10	30	75	105
11	65	88	153
12	54	18	72
13	55	42	97
14	60	32	92
15	13	22	35
16	56	25	81
17	99	6	105
18	60	12	72
19	13	12	25
20	56	56	112
21	96	77	173
22	55	44	99
23	44	66	110
24	33	18	51
25	45	10	55
26	85	8	93
27	58	6	64
28	99	25	124
29	25	47	72
30	26	26	52

C

#			
1	56	65	121
2	66	23	89
3	41	60	101
4	45	26	71
5	12	14	26
6	65	44	109
7	13	65	78
8	55	63	118
9	96	74	170
10	85	85	170
11	30	88	118
12	30	36	66
13	25	54	79
14	30	26	56
15	65	85	150
16	54	55	109
17	47	92	139
18	60	75	135
19	13	76	89
20	56	65	121
21	96	66	162
22	55	49	104
23	44	85	129
24	22	91	113
25	45	21	66
26	85	26	111
27	58	26	84
28	99	52	151
29	90	58	148
30	76	56	132

D

#			
1	33	56	89
2	30	66	96
3	25	45	70
4	30	45	75
5	65	12	77
6	66	65	131
7	47	13	60
8	60	35	95
9	13	96	109
10	55	88	143
11	30	30	60
12	36	30	66
13	64	25	89
14	12	39	51
15	72	45	117
16	85	54	139
17	52	47	99
18	39	60	99
19	55	34	89
20	51	56	107
21	58	96	154
22	32	55	87
23	88	45	133
24	75	22	97
25	76	25	101
26	85	9	94
27	25	58	83
28	99	5	104
29	45	90	135
30	88	76	164

ANSWERS PAGE 3

#	A			#	B			#	C			#	D		
1	25	-20	5	1	42	-3	39	1	66	-33	33	1	55	-10	45
2	25	-13	12	2	32	-6	26	2	63	-35	28	2	20	-8	12
3	52	-30	22	3	45	-8	37	3	39	-12	27	3	40	-14	26
4	85	-41	44	4	65	-4	61	4	45	-8	37	4	32	-13	19
5	25	-9	16	5	22	-3	19	5	22	-12	10	5	82	-9	73
6	66	-54	12	6	37	-15	22	6	65	-5	60	6	25	-18	7
7	52	-13	39	7	64	-32	32	7	44	-41	3	7	76	-15	61
8	39	-20	19	8	51	-41	10	8	55	-8	47	8	33	-14	19
9	74	-20	54	9	64	-50	14	9	96	-15	81	9	66	-44	22
10	79	-15	64	10	45	-6	39	10	85	-9	76	10	55	-12	43
11	95	-65	30	11	65	-3	62	11	30	-4	26	11	30	-16	14
12	32	-32	0	12	64	-23	41	12	30	-12	18	12	36	-16	20
13	65	-7	58	13	12	-10	2	13	25	-16	9	13	65	-45	20
14	12	-8	4	14	94	-4	90	14	88	-74	14	14	36	-12	24
15	72	-45	27	15	55	-47	8	15	65	-33	32	15	72	-33	39
16	85	-65	20	16	55	-30	25	16	98	-66	32	16	85	-25	60
17	25	-22	3	17	92	-53	39	17	47	-6	41	17	25	-12	13
18	10	-8	2	18	75	-23	52	18	60	-13	47	18	39	-10	29
19	11	-5	6	19	76	-55	21	19	13	-2	11	19	55	-24	31
20	51	-40	11	20	65	-5	60	20	56	-51	5	20	51	-8	43
21	64	-33	31	21	85	-70	15	21	96	-12	84	21	64	-12	52
22	48	-3	45	22	49	-19	30	22	55	-44	11	22	32	-5	27
23	85	-2	83	23	85	-14	71	23	44	-33	11	23	92	-41	51
24	48	-7	41	24	91	-5	86	24	22	-3	19	24	75	-8	67
25	12	-5	7	25	21	-6	15	25	45	-2	43	25	76	-15	61
26	88	-41	47	26	26	-1	25	26	85	-9	76	26	85	-9	76
27	21	-14	7	27	26	-12	14	27	58	-5	53	27	58	-4	54
28	99	-2	97	28	52	-7	45	28	99	-41	58	28	99	-12	87
29	47	-12	35	29	58	-47	11	29	90	-14	76	29	90	-23	67
30	76	-3	73	30	56	-26	30	30	76	-2	74	30	77	-6	71

ANSWERS PAGE 4

A

#			
1	44	-10	34
2	35	-8	27
3	45	-14	31
4	65	-13	52
5	25	-9	16
6	37	-4	33
7	64	-15	49
8	44	-14	30
9	64	-25	39
10	45	-12	33
11	66	-16	50
12	32	-16	16
13	65	-32	33
14	12	-12	0
15	72	-33	39
16	88	-25	63
17	25	-12	13
18	10	-8	2
19	55	-25	30
20	51	-8	43
21	66	-12	54
22	48	-8	40
23	85	-41	44
24	44	-8	36
25	12	-11	1
26	88	-9	79
27	32	-4	28
28	99	-14	85
29	47	-23	24
30	77	-6	71

B

#			
1	55	-33	22
2	55	-35	20
3	77	-12	65
4	85	-8	77
5	30	-14	16
6	44	-5	39
7	99	-88	11
8	88	-8	80
9	65	-15	50
10	78	-9	69
11	47	-7	40
12	64	-12	52
13	85	-16	69
14	94	-74	20
15	55	-21	34
16	87	-66	21
17	92	-6	86
18	75	-13	62
19	76	-7	69
20	77	-51	26
21	85	-15	70
22	21	-12	9
23	85	-33	52
24	91	-3	88
25	21	-2	19
26	55	-9	46
27	26	-8	18
28	52	-41	11
29	78	-14	64
30	56	-4	52

C

#			
1	66	-3	63
2	85	-13	72
3	25	-7	18
4	64	-4	60
5	55	-18	37
6	51	-15	36
7	77	-32	45
8	32	-21	11
9	92	-50	42
10	37	-6	31
11	76	-3	73
12	85	-23	62
13	58	-10	48
14	99	-9	90
15	90	-47	43
16	88	-30	58
17	96	-53	43
18	60	-5	55
19	65	-55	10
20	56	-5	51
21	77	-70	7
22	55	-35	20
23	44	-14	30
24	55	-5	50
25	45	-7	38
26	85	-15	70
27	75	-64	11
28	80	-7	73
29	90	-80	10
30	79	-26	53

D

#			
1	67	-20	47
2	55	-26	29
3	92	-30	62
4	75	-43	32
5	84	-9	75
6	65	-32	33
7	97	-13	84
8	33	-25	8
9	94	-20	74
10	55	-27	28
11	99	-65	34
12	36	-14	22
13	66	-8	58
14	36	-9	27
15	84	-45	39
16	85	-36	49
17	25	-22	3
18	39	-7	32
19	49	-5	44
20	51	-31	20
21	64	-33	31
22	91	-3	88
23	92	-74	18
24	83	-7	76
25	76	-4	72
26	73	-42	31
27	58	-14	44
28	99	-15	84
29	90	-75	15
30	82	-43	39

ANSWERS PAGE 5

A

#				
1	41	20	-18	43
2	16	22	-12	26
3	44	32	-15	61
4	33	16	-12	37
5	20	22	-20	22
6	45	41	-12	74
7	72	52	-32	92
8	62	36	-41	57
9	80	12	-8	84
10	14	75	-55	34
11	23	82	-12	93
12	36	14	-19	31
13	20	42	-19	43
14	66	32	-15	83
15	45	45	-62	28
16	55	65	-32	88
17	12	22	-5	29
18	41	10	-8	43
19	25	11	-6	30
20	55	51	-2	104
21	99	64	-45	118
22	32	45	-15	62
23	15	65	-15	65
24	13	18	-6	25
25	47	12	-6	53
26	88	3	-12	79
27	66	1	-15	52
28	82	98	-75	105
29	92	47	-33	106
30	24	76	-66	34

B

#				
1	15	85	-5	95
2	23	15	-12	26
3	60	45	-15	90
4	26	35	-8	53
5	14	23	-4	33
6	44	45	-12	77
7	55	74	-12	117
8	63	62	-25	100
9	74	82	-9	147
10	46	14	-54	6
11	14	38	-13	39
12	36	33	-20	49
13	14	20	-20	14
14	26	65	-16	75
15	85	45	-65	65
16	55	54	-32	77
17	92	12	-7	97
18	75	12	-8	79
19	76	13	-3	86
20	65	55	-2	118
21	32	96	-45	83
22	26	32	-14	44
23	85	12	-33	64
24	91	13	-4	100
25	21	47	-2	66
26	26	85	-9	102
27	26	66	-5	87
28	4	82	-42	44
29	18	91	-14	95
30	9	23	-2	30

C

#				
1	20	55	-4	71
2	25	22	-3	44
3	32	32	-6	58
4	16	18	-6	28
5	22	22	-5	39
6	44	41	-2	83
7	52	52	-6	98
8	36	37	-10	63
9	12	12	-9	15
10	79	75	-22	132
11	19	82	-2	99
12	32	18	-12	38
13	24	42	-6	60
14	12	32	-4	40
15	72	20	-3	89
16	32	65	-10	87
17	64	6	-24	46
18	10	11	-13	8
19	11	11	-2	20
20	51	55	-51	55
21	64	64	-12	116
22	48	45	-44	49
23	65	66	-33	98
24	18	18	-3	33
25	12	10	-2	20
26	32	3	-9	26
27	35	1	-5	31
28	98	99	-41	156
29	47	47	-14	80
30	76	76	-2	150

ANSWERS PAGE 6

A

1	52	74	-18	108
2	16	52	-12	56
3	33	82	-15	100
4	33	16	-12	37
5	20	38	-13	45
6	62	52	-12	102
7	72	21	-32	61
8	62	65	-32	95
9	80	40	-8	112
10	16	54	-55	15
11	23	12	-9	26
12	36	8	-19	25
13	44	13	-19	38
14	66	52	-15	103
15	45	96	-52	89
16	55	12	-32	35
17	44	8	-5	47
18	41	6	-8	39
19	25	47	-6	66
20	55	85	-2	138
21	99	66	-33	132
22	32	20	-15	37
23	15	91	-15	91
24	13	16	-6	23
25	47	12	-4	55
26	99	3	-12	90
27	66	3	-15	54
28	82	98	-13	167
29	92	47	-33	106
30	24	20	-23	21

B

1	22	85	-5	102
2	23	82	-12	93
3	60	12	-16	56
4	16	12	-8	20
5	22	47	-4	65
6	20	85	-9	96
7	52	16	-8	60
8	36	88	-25	99
9	12	91	-9	94
10	79	23	-44	58
11	19	17	-13	23
12	32	9	-20	21
13	24	9	-18	15
14	12	28	-12	28
15	16	47	-15	48
16	32	76	-5	103
17	64	22	-20	66
18	10	55	-12	53
19	80	13	-32	61
20	14	30	-41	3
21	23	96	-6	113
22	36	32	-55	13
23	20	11	-12	19
24	66	11	-4	73
25	45	47	-16	76
26	55	55	-9	101
27	77	66	-25	118
28	4	82	-42	44
29	18	36	-14	40
30	9	39	-2	46

C

1	10	55	-4	61
2	10	22	-3	29
3	32	36	-6	62
4	16	18	-3	31
5	22	22	-20	24
6	44	20	-12	52
7	64	50	-15	99
8	20	37	-12	45
9	11	12	-20	3
10	51	68	-12	107
11	64	55	-32	87
12	95	18	-41	72
13	85	42	-8	119
14	91	13	-55	49
15	21	5	-3	23
16	26	60	-10	76
17	28	6	-24	10
18	4	20	-13	11
19	11	41	-2	50
20	60	55	-51	64
21	64	64	-12	116
22	48	8	-44	12
23	65	76	-33	108
24	18	55	-3	70
25	99	10	-2	107
26	32	6	-9	29
27	35	12	-5	42
28	98	41	-41	98
29	47	20	-14	53
30	76	88	-2	162

ANSWERS PAGE 7

A

#				
1	88	74	-3	159
2	15	52	-3	64
3	48	82	-6	124
4	35	16	-7	44
5	25	38	-5	58
6	45	55	-2	98
7	75	21	-6	90
8	62	25	-12	75
9	82	40	-9	113
10	14	54	-22	46
11	39	12	-15	36
12	33	8	-12	29
13	20	15	-6	29
14	65	52	-10	107
15	47	96	-3	140
16	54	12	-10	56
17	18	9	-24	3
18	12	6	-13	5
19	46	47	-2	91
20	55	85	-51	89
21	96	67	-12	151
22	32	20	-44	8
23	12	91	-33	70
24	13	16	-14	15
25	88	12	-2	98
26	85	3	-20	68
27	66	12	-5	73
28	82	98	-44	136
29	85	47	-14	118
30	23	26	-12	37

B

#				
1	66	85	-15	136
2	85	88	-12	161
3	25	12	-15	22
4	65	12	-18	59
5	55	47	-20	82
6	51	95	-13	133
7	85	16	-20	81
8	32	88	-41	79
9	92	99	-7	184
10	37	23	-45	15
11	88	17	-13	92
12	85	15	-15	85
13	58	9	-22	45
14	99	36	-15	120
15	90	47	-65	72
16	99	76	-13	162
17	96	22	-7	111
18	60	56	-6	110
19	65	15	-9	71
20	56	30	-4	82
21	87	96	-26	157
22	55	32	-12	75
23	44	15	-15	44
24	55	11	-7	59
25	46	47	-4	89
26	85	60	-12	133
27	85	66	-16	135
28	80	82	-44	118
29	90	37	-33	94
30	79	42	-55	66

C

#				
1	55	55	-6	104
2	30	25	-8	47
3	25	36	-7	54
4	33	18	-4	47
5	65	40	-12	93
6	66	20	-12	74
7	47	55	-22	80
8	60	37	-42	55
9	45	12	-30	27
10	55	72	-6	121
11	30	55	-3	82
12	36	18	-33	21
13	64	42	-10	96
14	24	13	-6	31
15	72	5	-35	42
16	85	60	-30	115
17	55	6	-42	19
18	39	25	-23	41
19	55	41	-55	41
20	51	55	-42	64
21	65	64	-12	117
22	32	28	-5	55
23	88	76	-72	92
24	85	55	-18	122
25	76	12	-12	76
26	85	6	-15	76
27	25	12	-9	28
28	88	52	-9	131
29	45	20	-7	58
30	75	88	-8	155

ANSWERS PAGE 8

A

#					
1	10	20	-15	-5	10
2	30	22	-12	-3	37
3	40	32	-15	-2	55
4	5	16	-14	-4	3
5	10	22	-20	-5	7
6	30	41	-13	-1	57
7	10	52	-30	-5	27
8	30	36	-41	-10	15
9	10	12	-9	-9	4
10	30	78	-54	-20	34
11	10	19	-13	-2	14
12	30	32	-20	-12	30
13	10	24	-20	-5	9
14	30	6	-15	-4	17
15	10	72	-65	-3	14
16	30	32	-32	-17	13
17	42	6	-7	-24	17
18	30	10	-8	-13	19
19	10	11	-3	-4	14
20	30	51	-2	-51	28
21	10	64	-45	-12	17
22	30	45	-12	-40	23
23	10	65	-15	-33	27
24	30	18	-6	-3	39
25	10	12	-4	-2	16
26	30	3	-12	-7	14
27	47	1	-15	-5	28
28	30	98	-74	-41	13
29	10	47	-33	-14	10
30	30	76	-66	-2	38

B

#					
1	15	-3	63	-5	70
2	22	-6	15	-3	28
3	60	-8	42	-2	92
4	26	-4	35	-4	53
5	12	-3	27	-5	31
6	44	-12	45	-1	76
7	52	-32	65	-5	80
8	63	-41	75	-10	87
9	74	-50	82	-9	97
10	21	-5	14	-20	10
11	12	-3	38	-2	45
12	36	-23	32	-12	33
13	14	-10	20	-5	19
14	25	-7	65	-4	79
15	85	-47	41	-3	76
16	52	-30	54	-17	59
17	92	-53	12	-24	27
18	74	-23	9	-13	47
19	76	-55	13	-4	30
20	65	-42	54	-51	26
21	32	-14	96	-12	102
22	25	-6	32	-40	11
23	85	-70	9	-6	18
24	91	-19	2	-3	71
25	21	-14	47	-2	52
26	25	-3	85	-7	100
27	26	-6	61	-5	76
28	2	-1	82	-41	42
29	18	-14	90	-14	80
30	9	-7	23	-2	23

C

#					
1	20	20	-10	-5	25
2	65	22	-8	-12	67
3	41	32	-13	-15	45
4	54	16	-14	-6	50
5	12	22	-7	-4	23
6	9	41	-8	-12	30
7	13	52	-3	-15	47
8	54	36	-2	-74	14
9	96	12	-45	-9	54
10	30	75	-12	-54	39
11	10	82	-15	-13	64
12	30	14	-6	-20	18
13	10	42	-4	-20	28
14	30	32	-12	-15	35
15	65	20	-32	-45	8
16	54	65	-35	-32	52
17	45	6	-12	-7	32
18	60	10	-7	-8	55
19	13	11	-12	-3	9
20	54	51	-5	-2	98
21	96	64	-41	-45	74
22	32	45	-8	-12	57
23	9	65	-15	-33	26
24	2	18	-6	-3	11
25	47	12	-4	-2	53
26	85	3	-12	-7	69
27	61	1	-15	-5	42
28	82	98	-74	-41	65
29	90	47	-33	-14	90
30	23	76	-66	-2	31

ANSWERS PAGE 9

A

#					
1	85	20	-15	-5	85
2	15	22	-12	-3	22
3	45	32	-15	-2	60
4	35	16	-14	-4	33
5	23	22	-20	-5	20
6	45	41	-13	-1	72
7	74	52	-30	-5	91
8	62	36	-41	-10	47
9	82	12	-9	-9	76
10	14	75	-54	-20	15
11	38	82	-13	-2	105
12	33	14	-20	-12	15
13	20	42	-20	-5	37
14	65	32	-15	-4	78
15	45	45	-65	-3	22
16	54	65	-32	-17	70
17	12	22	-7	-24	3
18	12	10	-8	-13	1
19	13	11	-3	-4	17
20	55	51	-2	-51	53
21	96	64	-45	-12	103
22	32	45	-12	-40	25
23	12	65	-15	-33	29
24	13	18	-6	-3	22
25	47	12	-4	-2	53
26	85	3	-12	-7	69
27	66	1	-15	-5	47
28	82	98	-74	-41	65
29	91	47	-33	-14	91
30	23	76	-66	-2	31

B

#					
1	15	-3	20	-5	27
2	23	-6	25	-12	30
3	60	-8	32	-15	69
4	26	-4	16	-8	30
5	14	-3	22	-4	29
6	44	-15	44	-12	61
7	55	-32	52	-12	63
8	63	-41	36	-25	33
9	74	-50	12	-9	27
10	21	-6	79	-54	40
11	14	-3	19	-13	17
12	36	-23	32	-20	25
13	14	-10	24	-20	8
14	26	-4	12	-16	18
15	85	-47	72	-65	45
16	55	-30	32	-32	25
17	92	-53	6	-7	38
18	75	-23	10	-8	54
19	76	-55	11	-3	29
20	65	-42	51	-2	72
21	32	-14	64	-45	37
22	26	-5	48	-14	55
23	85	-70	65	-33	47
24	91	-19	18	-4	86
25	21	-14	12	-2	17
26	26	-5	3	-9	15
27	26	-6	1	-5	16
28	4	-1	98	-42	59
29	18	-12	47	-14	39
30	9	-7	76	-2	76

C

#					
1	21	55	-10	-4	62
2	66	22	-8	-3	77
3	41	32	-14	-6	53
4	45	18	-14	-6	43
5	12	22	-9	-5	20
6	9	41	-8	-2	40
7	13	52	-3	-6	56
8	55	37	-2	-10	80
9	96	12	-44	-9	55
10	31	75	-12	-22	72
11	10	82	-16	-2	74
12	30	18	-6	-12	30
13	12	42	-4	-6	44
14	30	32	-12	-4	46
15	65	20	-33	-3	49
16	54	65	-35	-10	74
17	47	6	-12	-24	17
18	60	11	-8	-13	50
19	13	11	-12	-2	10
20	54	55	-5	-51	53
21	96	64	-41	-12	107
22	32	45	-8	-44	25
23	11	66	-15	-33	29
24	2	18	-9	-3	8
25	45	10	-4	-2	49
26	85	3	-12	-9	67
27	58	1	-16	-5	38
28	82	99	-74	-41	66
29	90	47	-33	-14	90
30	25	76	-66	-2	33

A

#					
1	30	20	-15	-4	31
2	65	22	-15	-3	69
3	41	25	-15	-5	46
4	54	16	-23	-4	43
5	14	22	-20	-5	11
6	9	44	-14	-1	38
7	16	52	-30	-7	31
8	54	74	-39	-10	79
9	99	12	-9	-9	93
10	30	77	-54	-20	33
11	12	82	-13	-12	69
12	30	15	-21	-12	12
13	10	42	-20	-5	27
14	66	32	-12	-4	82
15	65	52	-65	-3	49
16	54	65	-25	-17	77
17	52	22	-7	-24	43
18	60	12	-9	-13	50
19	13	11	-3	-4	17
20	54	44	-6	-51	41
21	78	64	-45	-15	82
22	60	32	-15	-40	37
23	9	65	-15	-33	26
24	2	25	-6	-3	18
25	52	12	-4	-2	58
26	85	3	-12	-11	65
27	61	1	-15	-5	42
28	82	88	-74	-41	55
29	99	47	-33	-16	97
30	23	80	-66	-3	34

B

#					
1	40	-5	40	-5	70
2	22	-15	25	-10	22
3	32	-32	25	-12	13
4	16	-8	16	-8	16
5	41	-5	22	-4	54
6	55	-25	26	-12	44
7	52	-40	53	-14	51
8	36	-23	33	-25	21
9	12	-11	15	-9	7
10	75	-45	70	-54	46
11	87	-11	20	-13	83
12	63	-27	33	-20	49
13	42	-20	13	-20	15
14	32	-18	54	-16	52
15	84	-74	72	-66	16
16	65	-25	60	-32	68
17	16	-8	20	-7	21
18	10	-8	15	-8	9
19	11	-5	12	-3	15
20	66	-2	12	-2	74
21	64	-20	64	-45	63
22	45	-14	24	-14	41
23	65	-25	66	-33	73
24	18	-9	18	-4	23
25	24	-2	21	-2	41
26	35	-10	9	-9	25
27	36	-6	3	-5	28
28	98	-3	98	-42	151
29	47	-14	54	-14	73
30	45	-7	25	-2	61

C

#					
1	36	55	-4	-15	72
2	25	32	-3	-20	34
3	56	32	-6	-15	67
4	16	19	-6	-13	16
5	62	22	-20	-20	44
6	25	42	-12	-13	42
7	52	52	-17	-30	57
8	33	52	-12	-32	41
9	12	35	-20	-9	18
10	74	75	-10	-54	85
11	19	88	-32	-13	62
12	52	18	-42	-20	8
13	32	42	-8	-20	46
14	54	32	-45	-15	26
15	72	20	-3	-12	77
16	60	55	-10	-32	73
17	18	18	-24	-7	5
18	15	11	-13	-8	5
19	12	11	-2	-3	18
20	12	65	-51	-2	24
21	64	25	-12	-45	32
22	48	45	-40	-12	41
23	66	66	-33	-9	90
24	18	27	-3	-6	36
25	88	10	-2	-4	92
26	18	3	-5	-12	4
27	20	8	-5	-15	8
28	98	99	-32	-74	91
29	36	47	-14	-51	18
30	35	25	-2	-33	25

ANSWERS PAGE 11

A

#						
1	88	20	-12	-2	-5	89
2	17	22	-12	-4	-3	20
3	44	33	-15	-4	-2	56
4	30	16	-16	-4	-4	22
5	30	22	-20	-9	-5	18
6	44	44	-13	-1	-1	73
7	77	52	-20	-5	-5	99
8	63	26	-41	-12	-10	26
9	82	12	-9	-12	-9	64
10	15	78	-45	-22	-20	6
11	33	82	-13	-3	-2	97
12	35	15	-15	-12	-12	11
13	25	42	-20	-5	-5	37
14	66	33	-15	-4	-4	76
15	44	45	-65	-6	-3	15
16	50	66	-16	-17	-17	66
17	40	22	-9	-24	-24	5
18	30	12	-9	-13	-13	7
19	78	11	-9	-4	-4	72
20	90	51	-12	-25	-51	53
21	95	64	-25	-12	-12	110
22	67	47	-12	-30	-40	32
23	46	65	-15	-33	-33	30
24	28	78	-7	-9	-3	87
25	82	12	-4	-2	-2	86
26	65	36	-12	-17	-7	65
27	46	6	-16	-5	-5	26
28	95	14	-44	-21	-41	3
29	52	55	-33	-33	-14	27
30	32	85	-55	-6	-2	54

B

#						
1	15	-6	40	-5	14	58
2	22	-6	25	-12	25	54
3	60	-7	33	-32	30	84
4	26	-4	16	-8	16	46
5	15	-12	22	-4	36	57
6	44	-15	25	-12	44	86
7	56	-22	52	-40	52	98
8	63	-41	33	-25	36	66
9	74	-30	12	-9	12	59
10	22	-6	70	-54	79	111
11	14	-3	19	-11	19	38
12	35	-23	33	-20	33	58
13	15	-10	24	-20	24	33
14	26	-6	54	-16	12	70
15	88	-47	72	-70	62	105
16	55	-30	60	-35	33	83
17	99	-42	6	-8	6	61
18	75	-23	12	-8	14	70
19	85	-50	12	-3	11	55
20	65	-42	12	-2	52	85
21	32	-12	64	-60	66	90
22	13	-5	48	-14	44	86
23	85	-70	66	-30	65	116
24	91	-18	18	-4	44	131
25	20	-12	22	-2	12	40
26	26	-9	9	-10	8	24
27	36	-9	3	-5	1	26
28	14	-9	98	-25	44	122
29	18	-6	25	-14	25	48
30	10	-8	25	-6	12	33

ANSWERS PAGE 12

A

1	20	20	-4	-2	20	54
2	65	22	-3	-4	22	102
3	41	33	-6	-4	32	96
4	54	16	-3	-4	16	79
5	12	22	-20	-9	22	27
6	9	44	-12	-1	41	81
7	13	52	-15	-5	52	97
8	54	26	-12	-12	36	92
9	96	12	-20	-12	12	88
10	30	78	-12	-22	75	149
11	10	82	-32	-3	82	139
12	30	33	-41	-12	14	24
13	10	42	-8	-5	42	81
14	30	33	-55	-4	32	36
15	65	45	-3	-6	20	121
16	54	66	-10	-17	65	158
17	45	22	-24	-24	6	25
18	60	12	-13	-13	10	56
19	13	11	-2	-4	11	29
20	54	51	-51	-25	51	80
21	96	64	-12	-12	64	200
22	60	47	-44	-30	45	78
23	9	65	-33	-33	65	73
24	2	78	-3	-9	18	86
25	47	12	-2	-2	12	67
26	85	36	-9	-17	3	98
27	61	6	-5	-5	1	58
28	82	14	-41	-21	98	132
29	90	55	-14	-33	47	145
30	23	85	-2	-6	76	176

B

1	41	-15	40	-5	14	75
2	16	-12	25	-12	25	42
3	44	-15	33	-32	30	60
4	33	-14	16	-8	16	43
5	20	-20	22	-4	36	54
6	45	-13	25	-26	44	75
7	72	-30	52	-40	52	106
8	62	-41	33	-25	36	65
9	80	-9	12	-9	12	86
10	64	-54	70	-54	79	105
11	23	-13	19	-11	19	37
12	36	-20	33	-38	33	44
13	36	-20	22	-20	24	42
14	66	-15	54	-16	12	101
15	45	-3	72	-70	62	106
16	55	-32	60	-35	33	81
17	85	-7	6	-8	6	82
18	41	-8	12	-8	14	51
19	25	-3	12	-3	11	42
20	55	-2	12	-2	52	115
21	99	-45	64	-60	66	124
22	32	-12	48	-14	44	98
23	16	-15	66	-30	65	102
24	13	-6	18	-4	44	65
25	47	-4	22	-2	12	75
26	88	-12	9	-10	8	83
27	66	-15	3	-5	1	50
28	82	-74	98	-25	44	125
29	92	-33	25	-14	25	95
30	99	-66	25	-6	12	64

ANSWERS PAGE 13

A

1	102	242	-3	-200	20	161
2	22	12	-6	-4	22	46
3	123	62	-8	-4	32	205
4	18	121	-100	-4	16	51
5	22	6	-3	-9	22	38
6	441	14	-15	-210	41	271
7	225	11	-32	-5	52	251
8	550	111	-41	-221	36	435
9	12	66	-50	-12	12	28
10	321	44	-6	-22	75	412
11	82	65	-3	-3	82	223
12	18	623	-23	-12	14	620
13	42	42	-10	-22	42	94
14	251	33	-4	-4	32	308
15	20	45	-47	-6	20	32
16	65	412	-30	-17	65	495
17	333	22	-53	-120	6	188
18	412	12	-23	-13	10	398
19	11	223	-55	-4	11	186
20	550	51	-400	-25	51	227
21	213	821	-70	-300	64	728
22	45	90	-19	-60	45	101
23	166	65	-121	-33	65	142
24	18	78	-5	-9	18	100
25	362	12	-6	-2	12	378
26	875	36	-418	-17	3	479
27	365	6	-12	-5	1	355
28	999	14	-555	-21	98	535
29	400	55	-47	-33	47	422
30	76	288	-200	-6	76	234

B

1	300	-15	40	-10	14	329
2	22	-12	520	-8	25	547
3	152	-140	33	-14	30	61
4	36	-14	160	-13	16	185
5	220	-20	22	-9	36	249
6	75	-13	125	-18	44	213
7	882	-255	52	-15	52	716
8	140	-41	33	-44	36	124
9	425	-250	12	-44	12	155
10	632	-155	70	-120	79	506
11	120	-13	19	-16	19	129
12	65	-20	33	-16	33	95
13	600	-200	22	-310	24	136
14	100	-15	54	-12	12	139
15	110	-35	72	-33	62	176
16	665	-32	60	-150	33	576
17	452	-200	6	-120	6	144
18	210	-8	12	-10	14	218
19	110	-3	130	-24	11	224
20	230	-25	12	-8	52	261
21	64	-45	64	-12	66	137
22	125	-12	233	-155	44	235
23	610	-250	66	-200	65	291
24	180	-80	18	-8	44	154
25	120	-4	22	-15	12	135
26	125	-25	9	-9	8	108
27	225	-106	3	-40	1	83
28	99	-25	188	-12	44	294
29	850	-450	25	-23	25	427
30	320	-120	25	-6	12	231

ANSWERS PAGE 14

A

#						
1	120	242	-3	-12	20	367
2	222	100	-120	-4	22	220
3	133	62	-8	-25	32	194
4	16	121	-100	-4	16	49
5	250	6	-120	-9	22	149
6	44	320	-15	-120	41	270
7	520	225	-365	-5	52	427
8	26	111	-41	-12	36	120
9	120	66	-50	-120	12	28
10	780	44	-400	-22	75	477
11	82	665	-3	-3	82	823
12	330	623	-23	-12	14	932
13	42	250	-25	-5	42	304
14	124	33	-4	-24	32	161
15	450	45	-200	-6	20	309
16	66	412	-30	-17	65	496
17	222	22	-30	-24	6	196
18	12	444	-23	-44	10	399
19	110	223	-55	-4	11	285
20	51	515	-400	-25	51	192
21	132	821	-70	-520	64	427
22	165	90	-19	-30	45	251
23	65	241	-121	-14	65	236
24	136	78	-5	-9	18	218
25	12	444	-6	-2	12	460
26	950	555	-418	-600	3	490
27	963	6	-12	-540	1	418
28	652	100	-555	-21	98	274
29	55	200	-47	-33	47	222
30	425	288	-200	-6	76	583

B

#						
1	240	-6	120	-10	14	358
2	125	-6	420	-8	25	556
3	130	-7	55	-14	30	194
4	160	-4	160	-13	16	319
5	360	-12	140	-145	36	379
6	440	-15	125	-250	44	344
7	225	-200	52	-15	52	114
8	635	-410	33	-44	36	250
9	127	-30	12	-44	12	77
10	790	-650	70	-120	79	169
11	250	-54	19	-16	19	218
12	330	-23	33	-16	33	357
13	240	-110	200	-310	24	44
14	120	-60	54	-12	12	114
15	665	-470	72	-33	62	296
16	352	-300	500	-150	33	435
17	362	-320	600	-120	6	528
18	145	-23	12	-10	14	138
19	110	-50	130	-24	11	177
20	520	-420	12	-8	52	156
21	660	-500	64	-12	66	278
22	415	-350	233	-155	44	187
23	254	-145	66	-20	65	220
24	400	-200	18	-8	44	254
25	120	-25	22	-15	12	114
26	880	-741	9	-9	8	147
27	199	-99	3	-40	1	64
28	420	-145	188	-12	44	495
29	250	-125	25	-23	25	152
30	150	-50	25	-6	12	131

BLANK SHEETS FOR WORK PAGE ANSWERS

Page / Column		Page / Column		Page / Column		Page / Column		Page / Column	
1		1		1		1		1	
2		2		2		2		2	
3		3		3		3		3	
4		4		4		4		4	
5		5		5		5		5	
6		6		6		6		6	
7		7		7		7		7	
8		8		8		8		8	
9		9		9		9		9	
10		10		10		10		10	
11		11		11		11		11	
12		12		12		12		12	
13		13		13		13		13	
14		14		14		14		14	
15		15		15		15		15	
16		16		16		16		16	
17		17		17		17		17	
18		18		18		18		18	
19		19		19		19		19	
20		20		20		20		20	
21		21		21		21		21	
22		22		22		22		22	
23		23		23		23		23	
24		24		24		24		24	
25		25		25		25		25	
26		26		26		26		26	
27		27		27		27		27	
28		28		28		28		28	
29		29		29		29		29	
30		30		30		30		30	

WRITE THE ANSWERS FOR THE WORK PAGES

Page / Column		Page / Column		Page / Column		Page / Column		Page / Column	
1		1		1		1		1	
2		2		2		2		2	
3		3		3		3		3	
4		4		4		4		4	
5		5		5		5		5	
6		6		6		6		6	
7		7		7		7		7	
8		8		8		8		8	
9		9		9		9		9	
10		10		10		10		10	
11		11		11		11		11	
12		12		12		12		12	
13		13		13		13		13	
14		14		14		14		14	
15		15		15		15		15	
16		16		16		16		16	
17		17		17		17		17	
18		18		18		18		18	
19		19		19		19		19	
20		20		20		20		20	
21		21		21		21		21	
22		22		22		22		22	
23		23		23		23		23	
24		24		24		24		24	
25		25		25		25		25	
26		26		26		26		26	
27		27		27		27		27	
28		28		28		28		28	
29		29		29		29		29	
30		30		30		30		30	

Page / Column		Page / Column		Page / Column		Page / Column		Page / Column	
1		1		1		1		1	
2		2		2		2		2	
3		3		3		3		3	
4		4		4		4		4	
5		5		5		5		5	
6		6		6		6		6	
7		7		7		7		7	
8		8		8		8		8	
9		9		9		9		9	
10		10		10		10		10	
11		11		11		11		11	
12		12		12		12		12	
13		13		13		13		13	
14		14		14		14		14	
15		15		15		15		15	
16		16		16		16		16	
17		17		17		17		17	
18		18		18		18		18	
19		19		19		19		19	
20		20		20		20		20	
21		21		21		21		21	
22		22		22		22		22	
23		23		23		23		23	
24		24		24		24		24	
25		25		25		25		25	
26		26		26		26		26	
27		27		27		27		27	
28		28		28		28		28	
29		29		29		29		29	
30		30		30		30		30	

WRITE THE ANSWERS FOR THE WORK PAGES

Page / Column		Page / Column		Page / Column		Page / Column		Page / Column	
1		1		1		1		1	
2		2		2		2		2	
3		3		3		3		3	
4		4		4		4		4	
5		5		5		5		5	
6		6		6		6		6	
7		7		7		7		7	
8		8		8		8		8	
9		9		9		9		9	
10		10		10		10		10	
11		11		11		11		11	
12		12		12		12		12	
13		13		13		13		13	
14		14		14		14		14	
15		15		15		15		15	
16		16		16		16		16	
17		17		17		17		17	
18		18		18		18		18	
19		19		19		19		19	
20		20		20		20		20	
21		21		21		21		21	
22		22		22		22		22	
23		23		23		23		23	
24		24		24		24		24	
25		25		25		25		25	
26		26		26		26		26	
27		27		27		27		27	
28		28		28		28		28	
29		29		29		29		29	
30		30		30		30		30	

Page / Column		Page / Column		Page / Column		Page / Column		Page / Column	
1		1		1		1		1	
2		2		2		2		2	
3		3		3		3		3	
4		4		4		4		4	
5		5		5		5		5	
6		6		6		6		6	
7		7		7		7		7	
8		8		8		8		8	
9		9		9		9		9	
10		10		10		10		10	
11		11		11		11		11	
12		12		12		12		12	
13		13		13		13		13	
14		14		14		14		14	
15		15		15		15		15	
16		16		16		16		16	
17		17		17		17		17	
18		18		18		18		18	
19		19		19		19		19	
20		20		20		20		20	
21		21		21		21		21	
22		22		22		22		22	
23		23		23		23		23	
24		24		24		24		24	
25		25		25		25		25	
26		26		26		26		26	
27		27		27		27		27	
28		28		28		28		28	
29		29		29		29		29	
30		30		30		30		30	

WRITE THE ANSWERS FOR THE WORK PAGES

Page / Column		Page / Column		Page / Column		Page / Column		Page / Column	
1		1		1		1		1	
2		2		2		2		2	
3		3		3		3		3	
4		4		4		4		4	
5		5		5		5		5	
6		6		6		6		6	
7		7		7		7		7	
8		8		8		8		8	
9		9		9		9		9	
10		10		10		10		10	
11		11		11		11		11	
12		12		12		12		12	
13		13		13		13		13	
14		14		14		14		14	
15		15		15		15		15	
16		16		16		16		16	
17		17		17		17		17	
18		18		18		18		18	
19		19		19		19		19	
20		20		20		20		20	
21		21		21		21		21	
22		22		22		22		22	
23		23		23		23		23	
24		24		24		24		24	
25		25		25		25		25	
26		26		26		26		26	
27		27		27		27		27	
28		28		28		28		28	
29		29		29		29		29	
30		30		30		30		30	

Page / Column		Page / Column		Page / Column		Page / Column		Page / Column	
1		1		1		1		1	
2		2		2		2		2	
3		3		3		3		3	
4		4		4		4		4	
5		5		5		5		5	
6		6		6		6		6	
7		7		7		7		7	
8		8		8		8		8	
9		9		9		9		9	
10		10		10		10		10	
11		11		11		11		11	
12		12		12		12		12	
13		13		13		13		13	
14		14		14		14		14	
15		15		15		15		15	
16		16		16		16		16	
17		17		17		17		17	
18		18		18		18		18	
19		19		19		19		19	
20		20		20		20		20	
21		21		21		21		21	
22		22		22		22		22	
23		23		23		23		23	
24		24		24		24		24	
25		25		25		25		25	
26		26		26		26		26	
27		27		27		27		27	
28		28		28		28		28	
29		29		29		29		29	
30		30		30		30		30	

WRITE THE ANSWERS FOR THE WORK PAGES

Page / Column		Page / Column		Page / Column		Page / Column		Page / Column	
1		1		1		1		1	
2		2		2		2		2	
3		3		3		3		3	
4		4		4		4		4	
5		5		5		5		5	
6		6		6		6		6	
7		7		7		7		7	
8		8		8		8		8	
9		9		9		9		9	
10		10		10		10		10	
11		11		11		11		11	
12		12		12		12		12	
13		13		13		13		13	
14		14		14		14		14	
15		15		15		15		15	
16		16		16		16		16	
17		17		17		17		17	
18		18		18		18		18	
19		19		19		19		19	
20		20		20		20		20	
21		21		21		21		21	
22		22		22		22		22	
23		23		23		23		23	
24		24		24		24		24	
25		25		25		25		25	
26		26		26		26		26	
27		27		27		27		27	
28		28		28		28		28	
29		29		29		29		29	
30		30		30		30		30	

Page / Column		Page / Column		Page / Column		Page / Column		Page / Column	
1		1		1		1		1	
2		2		2		2		2	
3		3		3		3		3	
4		4		4		4		4	
5		5		5		5		5	
6		6		6		6		6	
7		7		7		7		7	
8		8		8		8		8	
9		9		9		9		9	
10		10		10		10		10	
11		11		11		11		11	
12		12		12		12		12	
13		13		13		13		13	
14		14		14		14		14	
15		15		15		15		15	
16		16		16		16		16	
17		17		17		17		17	
18		18		18		18		18	
19		19		19		19		19	
20		20		20		20		20	
21		21		21		21		21	
22		22		22		22		22	
23		23		23		23		23	
24		24		24		24		24	
25		25		25		25		25	
26		26		26		26		26	
27		27		27		27		27	
28		28		28		28		28	
29		29		29		29		29	
30		30		30		30		30	

WRITE THE ANSWERS FOR THE WORK PAGES

Page / Column		Page / Column		Page / Column		Page / Column		Page / Column	
1		1		1		1		1	
2		2		2		2		2	
3		3		3		3		3	
4		4		4		4		4	
5		5		5		5		5	
6		6		6		6		6	
7		7		7		7		7	
8		8		8		8		8	
9		9		9		9		9	
10		10		10		10		10	
11		11		11		11		11	
12		12		12		12		12	
13		13		13		13		13	
14		14		14		14		14	
15		15		15		15		15	
16		16		16		16		16	
17		17		17		17		17	
18		18		18		18		18	
19		19		19		19		19	
20		20		20		20		20	
21		21		21		21		21	
22		22		22		22		22	
23		23		23		23		23	
24		24		24		24		24	
25		25		25		25		25	
26		26		26		26		26	
27		27		27		27		27	
28		28		28		28		28	
29		29		29		29		29	
30		30		30		30		30	

WRITE THE ANSWERS FOR THE WORK PAGES

Page / Column		Page / Column		Page / Column		Page / Column		Page / Column	
1		1		1		1		1	
2		2		2		2		2	
3		3		3		3		3	
4		4		4		4		4	
5		5		5		5		5	
6		6		6		6		6	
7		7		7		7		7	
8		8		8		8		8	
9		9		9		9		9	
10		10		10		10		10	
11		11		11		11		11	
12		12		12		12		12	
13		13		13		13		13	
14		14		14		14		14	
15		15		15		15		15	
16		16		16		16		16	
17		17		17		17		17	
18		18		18		18		18	
19		19		19		19		19	
20		20		20		20		20	
21		21		21		21		21	
22		22		22		22		22	
23		23		23		23		23	
24		24		24		24		24	
25		25		25		25		25	
26		26		26		26		26	
27		27		27		27		27	
28		28		28		28		28	
29		29		29		29		29	
30		30		30		30		30	

Printed in Great Britain
by Amazon